Industrial Electromagnetics Modelling

Developments in Electromagnetic Theory and Applications

J. Heading, *Managing Editor*

Industrial Electromagnetics Modelling

Proceedings of the POLYMODEL 6, the Sixth Annual Conference of the North East Polytechnics Mathematical Modelling and Computer Simulation Group, held at the Moat House Hotel, Newcastle upon Tyne, May 1983

Edited by

J. Caldwell and R. Bradley

Newcastle upon Tyne Polytechnic

1983 **Springer-Science+Business Media, B.V.**

Distributors

for the United States and Canada: Kluwer Boston, Inc., 190 Old Derby Street, Hingham, MA 02043, USA
for all other countries: Kluwer Academic Publishers Group, Distribution Center, P.O.Box 322, 3300 AH Dordrecht, The Netherlands

Library of Congress Cataloging in Publication Data

North East Polytechnics Mathematical Modelling and Computer Simulation Group.
    Conference (6th : 1983 : Newcastle upon tyne, Tyne and Wear) Industrial
    Electromagnetics Modelling.

    (Developments in Electromagnetic Theory and Applications : v. 1)
      1. Electromagnetism – Data processing – Congresses. 2. Electromagnetism –
Mathematical Models – Congresses. 3. Digital computer simulation –
Congresses. i. Caldwell, J. (Jim) ii. Bradley, R. iii. Title. iv. series.
QC670.N68   1983   621.3'0724   83-22001

ISBN 978-94-009-6919-3

ISBN 978-94-009-6919-3   ISBN 978-94-009-6917-9 (eBook)
DOI 10.1007/978-94-009-6917-9

Copyright

PREFACE

During the past few years the rapid development of computer technology has made high power computing facilities more readily
accessible to a greater proportion of our industrial and academic
community. This development coupled with the recent upsurge in
mathematical modelling and computer simulation has led to significant developments in electromagnetic field theory and its applications to industry. In view of such developments and the present
high interest to both academics and industry the theme chosen for
the Polymodel 6 Conference held at Newcastle upon Tyne in May 1983
was Industrial Electromagnetics Modelling.

To date the North East Polytechnics Mathematical Modelling and
Computer Simulation Group has organised five successful Polymodel.
conferences each with a different theme. The objectives of the
Polymodel group include the promotion of collaborative research
between Newcastle, Sunderland and Teesside Polytechnics and industry
in the areas of mathematical modelling and computer simulation.

The aim of the Polymodel 6 Conference was to call on and use the
modelling and computer simulation expertise of eminent academics
and industrialists who are deeply involved in the area of electromagnetics. These proceedings have a twofold purpose in that they
contain current analytical and numerical techniques relevant to
electromagnetic field problems and useful ideas on the modelling and
simulation techniques which are most appropriate. It was also felt
important to include implications of computer developments (both
hardware and software) on such work.

The book is conveniently devided up into the following three
sections:

1) Applications and modelling techniques
2) Simulation and computing techniques
3) Impact of computer development

Over a period of three days the conference took the form of six
sessions each of which consisted of an introductory talk by a
keynote speaker followed by related talks by other contributors
from industry and academic establishments.

The text book is particularly intended for academics and industrialists interested in recent developments in electromagnetic theory and

applications. This would include persons with a background in applied mathematics, electrical engineering or computer science. The book should also be of value to postgraduates researching in this area and to recent recruits to industry.

J CALDWELL & R BRADLEY
Conference Chairman & Secretary

CONTENTS

Preface   V

SESSION A - APPLICATIONS & MODELLING TECHNIQUES I

Keynote talk
- Electromagnetic Modelling of Electrical Machinery
for Design Applications
M V K Chari
General Electric Company, New York   3

A Detailed Lumped-Parameter Method for the Study of
Electrical Machines in Transient Stability Studies
L Haydock & D C MacDonald
Trent Polytechnic & Imperial College   15

The Electromagnetic Field of an Insulated Conductor in
a Slot
S G Mudge
Wolverhampton Polytechnic   27

SESSION B - APPLICATIONS & MODELLING TECHNIQUES II

Keynote talk
- Electromagnetic Modelling Techniques using Finite
Element Methods
Z J Cendes
Carnegie-Mellon University, Pittsburgh   41

Error Bounded Formulations in Electromagnetism
J Penman & J R Fraser
University of Aberdeen & McDermott Engineering Ltd.   53

SESSION C - APPLICATIONS & MODELLING TECHNIQUES III

Keynote talk
- Electromagnetic Modelling Techniques using Boundary
Element Methods
A Wexler
University of Manitoba   65

Magnetostatic Field Calculations associated with Thick
Solenoids with Iron Present
J Caldwell, A Zisserman & R Saunders
Newcastle Polytechnic & Sunderland Polytechnic   77

Periodic Solutions for certain Non-Linear Parabolic
Partial Differential Equations
G Gregory
NEI Parsons Ltd.   89

SESSION D - SIMULATION & COMPUTING TECHNIQUES I

Keynote talk
- CAD Aspects of Electromagnetic Field Calculations
E M Freeman
Imperial College   103

Simulation of Hysteresis in Ferromagnetic Materials
D O'Kelly
University of Bradford   115

Prediction of Stability Models using Frequency Response
Data
A H Whitfield
Loughborough University of Technology   125

SESSION E - SIMULATION & COMPUTING TECHNIQUES II

Keynote talk
- The Extraction of Engineering Information from a
Potential Solution to an Electromagnetic Field Problem
D A Lowther
McGill University, Montreal   143

Optimisation of the Magnetic Screening of Electromagnetic
Coils
A Zisserman, J Caldwell & D H Prothero
Sunderland Polytechnic, Newcastle Polytechnic &
International Research & Development Co. Ltd.   155

Performance Calculations for Devices with Permanent
Magnets
D Howe & W F Low
University of Sheffield   167

The Use of Complex Permeability for Steady State Non-
Linear Eddy-Current and Hysteresis Problems
A G Jack, M R Harris & P T Jowett
University of Newcastle upon Tyne & Ferranti Ltd.
Edinburgh   179

SESSION F - IMPACT OF COMPUTER DEVELOPMENT

Keynote talk
- Hardware-Dependence of Electromagnetics Software
P P Silvester
McGill University, Montreal   189

Techniques of Post-Processing for Electromagnetic Field
Solutions
M L Barton, I A Ince & J J Oravec
Westinghouse Electric Corporation, Pittsburgh   201

Computer Development at Sunderland Polytechnic - Has
it Proved to be a Periodic Function?
E Crompton & J Tindle
Sunderland Polytechnic   213

A Novel Approach to Computer Architecture
C Dennison
Perkin-Elmer Data Systems, Slough   229

List of Delegates   233

SESSION A

APPLICATIONS & MODELLING TECHNIQUES I

Chairman

D R TREECE

NEI Parsons

ELECTROMAGNETIC MODELING OF ELECTRICAL MACHINERY
FOR DESIGN APPLICATIONS

M.V.K. Chari
General Electric Company
Schenectady, NY 12301

ABSTRACT

The need for design optimization and operational reliability of electrical machinery necessitates accurate performance prediction at the design stage. This in turn requires the detailed evaluation of the magnetic field distribution in the machine geometry. The finite element method offers a stable numerical solution technique with a good deal of precision. This method requires the formulation of the partial differential equations in variational terms or by a weighted residual procedure. In the case of the former an energy related expression called a functional is minimized with respect to a set of trial solutions. This paper describes the application of the finite element method to electrical machine modeling taking into account iron saturation, eddy currents in conducting parts and the distributed nature of materials and sources. Some of the areas surveyed are utility generators, skin-effect in busbars, eddy current losses in transformers, and others.

1. INTRODUCTION

Modern electrical plant and machinery are required to operate at optimal efficiency with a high degree of reliability. These requirements coupled with a demand for economy on the size and weight of machinery necessitates accurate performance prediction at the design stage. Some of the performance indicators of interest to the designer are excitation requirements on no-load and full-load conditions, reactances, transient characteristics, short-circuit forces, iron and stray load losses, end-region fields, eddy current effects in conducting media, dielectric stresses in insulating parts, etc.

In order to evaluate these parameters with precision, one must predict accurately and in detail the electric and magnetic fields produced under various operating conditions. Classical analysis techniques [1] and analog methods [2] have not been found to be effective except for simplified geometrical shapes and boundary conditions. The need for improved modeling techniques was, therefore, recognized even at the early stages of the design art. With the advent of electronic computing machines, numerical methods have come to the fore as effective and accurate methods of determining the field distribution in electrical apparatus. The principal numerical methods currently in use fall into three categories; namely, (i) divided difference schemes

[3], (ii) integral equation techniques [4], and (iii) projective and variational formulations [5,6]. Many variants of the above and hybrid methods have also appeared in the research literature. All of these methods address the electromagnetic field problem in terms of a mathematical model, its solution, and manipulation of the results to obtain design parameters.

Modeling consists of formulating the physical problem in mathematical terms from Maxwell's equations and obtaining the necessary differential or integral expressions which describe the electromagnetic relationships. The solution phase involves finding an approximation to the true solution which satisfies these equations so as to minimize the error in a prescribed sense. Several approaches are possible for the choice of the solution approximation and the error norm.

In this paper, the method of modeling the field problem by partial differential equations in terms of potential functions, the discretization of the field region into finite elements, the construction of an approximate solution by means of interpolatory basis functions, the setting up of an error norm by variational methods, and minimizing the error in a least square sense to obtain the potential solution will be described in some detail.

The principal differential equation formulation considered in this paper for modeling purposes comprise (i) vector Poisson equation for linear and nonlinear magnetostatic problems, (ii) diffusion equation, representative of steady state or transient eddy current and skin-effect problems and (iii) scalar Poisson equation for electrostatic field problems. The geometries modeled include two-dimensional, axisymmetric and three-dimensional representations of the field region. A second order method of quasilinearizing the nonlinear equations resulting from finite element discretization and functional minimization is described and their solutions by efficient computational methods are discussed. These include the Newton-Raphson algorithm, variants of the Gaussian elimination technique and the preconditioned conjugate gradient method. Post-processing of the field solution results in terms of flux plots and parameters such as flux-densities, voltages, currents, etc., are illustrated.

The paper also presents applications of the analyses described, to a variety of engineering design problems. These range from no-load and load analyses of utility generators [7,8], skin-effect in busbars [9], eddy current losses in transformers [10], no-load and starting performance in induction machines [11], dc machine fields [12], permanent magnet device performance prediction [13], eddy current non-destructive evaluation [14] and electrostatic field applications [15], and others. Comparison of finite element analysis results with test values are presented wherever feasible.

4

2. THE TWO-DIMENSIONAL MAGNETOSTATIC FIELD PROBLEM

i. Nonlinear Poisson Equation and the energy related functional:

In many electrical machines applications, modeling of the cross-sectional geometry is considered adequate for design purposes, although such an assumption neglects end-effects which are well and truly three-dimensional. In addition, in a good number of cases, the effects due to eddy currents and circulating currents are neglected and emphasis is placed on magnetic saturation of the iron parts. Subject to these assumptions and that of a single valued B-H characteristic, the magnetic field problem in the device may be formulated in terms of the nonlinear Poisson equation

$$\text{curl } \nu \text{ curl } \bar{A} = -\bar{J} \qquad (1)$$

where ν : is the reciprocal permeability which is both position and field dependent

$\bar{A}$: is the vector potential function which has a 'z' directed component only

$\bar{J}$: is the source current density which is also 'z' directed.

In the finite element method based on a variational formulation, equation (1) is reformulated in terms of an energy related expression called the functional. An approximation to the field solution is then sought which makes the functional stationary, a condition both necessary and sufficient to obtain the correct solution in a least square sense. The conventional procedure of formulating the functional of equation (1) for 2-D problems by variational methods yields the expression

$$\mathscr{F} = \int \{\nu \mid \text{grad } A \mid^2 - 2\,\bar{A}.\bar{J}\,\} \, dV - \oint_s \nu(\,A \text{ grad } A).\,\overline{ds} \qquad (2)$$

Homogeneous boundary conditions of the Dirichlet or Neumann type are satisfied by setting the line integral in (2) to zero. This is one of the principal advantages of the finite element method, wherein no special prescription for imposing natural boundary conditions is necessary. Examination of equation (2) reveals that it has the dimensionality of energy and hence the name energy-related functional. Substituting for grad A, the flux-density $\bar{B}$, one obtains the nonlinear functional

$$\mathscr{F} = 2\iint [\int_0^{\bar{B}} \nu \bar{B}.d\bar{B} - \bar{J}.\bar{A}] \, dx \, dy - \oint_{\Gamma} \nu\, A\, \frac{\partial A}{\partial n}.d\Gamma \qquad (3)$$

5

ii. Finite Element Approximation

The finite element procedure consists of subdividing the field region into a discrete number of sub-regions or elements, and prescribing the approximate solution in each of the sub-divisions. This approximation is obtained as a linear combination of solution values at a set of nodes, usually the corners and interior points, weighted by interpolatory functions. The geometrical shape of the element and the approximate solution defined in it describe the finite element. A variety of element shapes are available to the analyst ranging from the simple planar triangular or quadrilateral elements to the more sophisticated curvilinear and isoparametric elements of the first and higher order interpolation. In a majority of cases, only function continuity at the nodes and edges of the elements is required.

Figure 1 shows first-order triangular finite elements which are used in discretizing the field region. The potential solution is defined in terms of interpolatory functions of the triangular geometry and its nodal values of potential. Thus,

$$A = \sum_{k=i}^{m} \zeta_k A_k \tag{4}$$

where ζ_i, A_i are the value of the shape function and the potential at the ith node respectively.

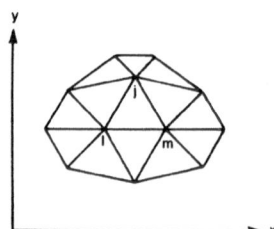

Figure 1. Discretization of the field region by triangular finite elements

Typically for triangular elements, the interpolatory functions are defined by Silvester's polynomial [5] in triple subscript notation as

$$\zeta_i^{(N)}(\xi) = \left(\frac{N\xi - i + 1}{i}\right) \zeta_{i-1}^{(N)}(\xi) \qquad \text{if } i > 0 \tag{5}$$

$$\zeta_o^{(N)}(\xi) = 1$$

$$\zeta_i^{(N)}(\xi) = 0 \qquad \text{if } i < 0 \tag{6}$$

where ξ's are the natural or area coordinates of the triangular region. The interpolation functions should satisfy the requirements; (i) that the function has a unit value at each node and is zero at the other nodes; (ii) sum of the interpolatory or shape functions must always equal unity. The solution surface for a typical second order triangular element is illustrated in Figure 2.

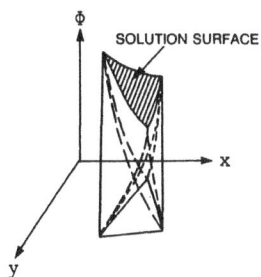

Figure 2. Illustration of the approximate solution
with a second order triangular element

iii. Magnetic Field Solution

The solution to the field problem is obtained by either minimizing the functional or by the weighted residual method generally known as the Galerkin criterion. In the former, the stationary value of the functional is found by setting its first derivative to zero with respect to each of the values of the approximate potential at each node. For a quadratic functional of the form of equation (3), the stationary value of the functional is also its minimum value. This procedure may be formally stated as

$$\frac{\partial F}{\partial A_K}\bigg|_{K=1,2,\ldots n} = 0 \tag{7}$$

The minimization is carried out over all the triangles of the region and the respective matrices are assembled to form the global matrix equation, the solution of which yields the unknown potential A.

The matrix equation may be formally stated as

$$[S]\ [A] = [T] \tag{8}$$

In the weighted residual procedure, one applies the Galerkin criterion to the differential equation (1) in terms of the interpolation functions ζ_i, ζ_j, ζ_m, and the expansion for $\bar{A}$ given by equation (4) such that

$$\int \zeta_i \cdot \text{curl } \nu \text{ curl} \sum \zeta_k A_K \, dV = 0 \tag{9}$$

7

It can be shown that the successive expansion of (9) over each of the nodes i,j,m of the finite element yields the same matrix equation as that obtained by functional minimization in (7). In the following sections, therefore, further reference to the weighted residual method will not be made.

iv. Quasi-linearization and solution of linear equations

Equation (8) is nonlinear, whose coefficient matrix is symmetric, sparse and band-structured. This equation represents the discretized version of the differential equation (1) obtained by a variational procedure. It is first quasi-linearized by the Newton-Raphson algorithm and the resulting set of linear equations is solved directly or indirectly. The reluctivity is updated in each iteration with respect to the B-H characteristic of the material. The N-R algorithm and the solution of the linear equations are described in reference [16] as follows.

The kth iterate of the vector potential yields the (k+1)th iterate by the recursive relationship

$$A^{k+1} = A^k - [J]^{-1} [S \; A^k - T] \qquad (10)$$

where [J] is the Jacobian matrix of partial derivatives of the iteration function [S A^k - T].

Silvester, Cabayan and Browne [17] applied the N-R algorithm to the functional formulation instead of the residual vector, such that

$$A^{k+1} = A^k - \left[\frac{\partial^2 \mathscr{F}}{\partial A_i \; \partial A_j}\right]^{-1} \cdot \left[\frac{\partial \mathscr{F}}{\partial A_i}\right] \qquad (11)$$

The above formulation has the attractive feature that it provides the theoretical foundation for high-order elements.

The set of linearized equations resulting from (10) is solved by Gaussian elimination technique with node re-numbering [18], profile storage [19] and wave front solution methods [20]. Recently, however, the pre-conditioned conjugate gradient method has been successfully applied [21] with utmost storage economy and fast execution time.

v. Magnetic flux-distribution in rotating electric machinery

Figure 3(a) shows the magnetic flux plot obtained by Minnich et al [8] for a 50 MVA turbine-generator on no-load. Comparison of test values and finite element results for the open -circuit characteristic is illustrated in Figure 3(b).

8

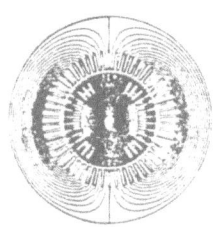

Figure 3(a). Flux plot of a turbine-
generator on no-load.

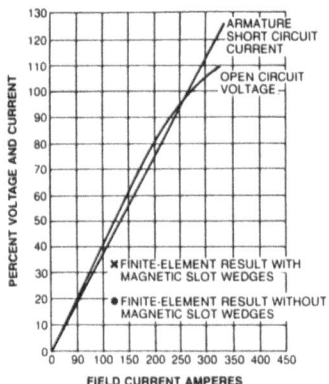

Figure 3(b). Comparison of finite element
calculation of open-circuit
voltage with test values.

During load operation of a rotating electric machine, the axes of
symmetry of the magnetic field depart a good deal from the polar and
interpolar axes. Under these conditions, the vector potential at any
point along the pole axis equals the magnitude at the corresponding
point a pole pitch away, but is of opposite sign to the former. This
is commonly known as the 'periodicity' condition which can be viewed
as an additional boundary condition. Thus, the coefficient matrix and
forcing function of equation (8) are modified such that the sign of
the off-diagonal terms corresponding to the points where the periodicity condition applies is reversed. This process is implemented by
means of a connection matrix described in reference [16]. Figure 4
illustrates the shift in the magnetic axis on load for a dc-generator
[12].

Minnich et al [8] applied the finite-element method for analyzing
the field distribution in a turbine-generator on steady state load
operation for various power-factors. A typical flux plot obtained for
a loaded generator is shown in Figure 5.

Since the time variation of flux is sinusoidal, the induced voltage relates linearly to flux-linkages and, hence, also to the vector
potential. The flux-linkages in the direct and quadrature axes are
calculated in terms of the vector potential, which should accurately
yield the direct and quadrature components of voltage. In general,
this is a two-step process. Brandl, Reichert, and Vogt [7] suggested
an alternative scheme employing the Newton-Raphson technique. In this
method, the constraint conditions on the terminal quantities calculated are imposed implicitly in the finite element formulation.

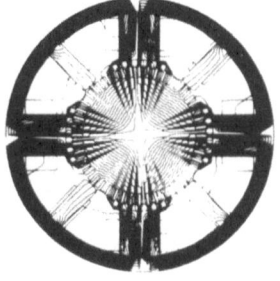

Figure 4. Magnetic flux
distribution in a
dc-generator on load.

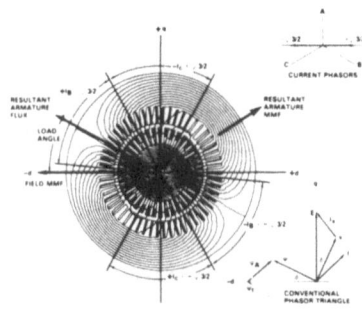

Figure 5. Flux distribution
on load of a 50 MVA
turbine-generator.

3. ANALYSIS OF THE EDDY CURRENT DIFFUSION PROBLEM

The first application of the finite element method to the analysis of eddy current fields in magnetotelluric problems was presented by Silvester and Haslam [22]. Chari [23] showed a similar formulation for linear eddy-current problems in magnetic structures. The governing partial differential equation and the functional formulation in two-dimensional Cartesian system may be expressed as

$$\nu\left(\frac{\partial^2 A}{\partial x^2} + \frac{\partial^2 A}{\partial y^2}\right) = \frac{\mathbf{i}j\omega A}{\rho} - J_s \tag{12}$$

where $\bar{A}$ is assumed to be a periodic function of time. The functional expression is given by

$$\mathscr{F} = \iint \frac{\nu}{2}\left\{\left(\frac{\partial A}{\partial x}\right)^2 + \left(\frac{\partial A}{\partial y}\right)^2\right\}\, dxdy + \frac{j\omega\mu}{2\rho}\iint A^2\, dxdy$$

$$-\iint J_s \cdot A\, dxdy \tag{13}$$

The minimization of the above functional for first-order triangular elements leads to a complex matrix equation of the form

$$[S]\,[A] + [C]\,[A] = [T] \tag{14}$$

The analysis was applied to semi-infinite and conducting slabs and the results were compared with test values, with good agreement. Another application of the method to a utility generator for obtaining its operational impedances was described in reference [24]. The flux plots obtained at 0.1 Hz are shown in Figure 6. The steady state diffusion equation solution of (14) was also applied to the nondestructive evaluation [14] of a crack in a bar of aluminum as illustrated in Figures 6(a) and 6(b).

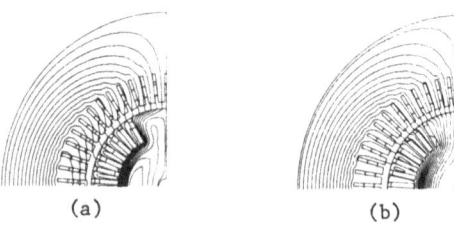

(a)   (b)

Figure 6. Flux Plots for 0.1 hertz d-axis current. Left: real component. Right: imaginary components. Skin depths: rotor, 2.59 inches, wedges, 16.7 inches.

Donea, Giuliani and Philippe [25] presented a useful analysis of the axisymmetric electromagnetic induction problem with a 'θ' directed vector potential A_θ (r,z). Foggia, Sabonnadière and Silvester [27] presented a solution of the saturated magnetic field problem in a linear induction motor. The partial differential equation was modified to include a velocity term 'V' so that

$$\frac{\partial}{\partial x}\left(\frac{1}{\mu}\frac{\partial A}{\partial x}\right) + \frac{\partial}{\partial y}\left(\frac{1}{\mu}\frac{\partial A}{\partial y}\right) = \frac{1}{\rho}\left(\frac{\partial A}{\partial t} + V\frac{\partial A}{\partial x}\right) \tag{15}$$

An alternative to the functional minimization process was presented using the weak Galerkin formulation with the result

$$\int_\Omega \left(\frac{\partial A}{\partial t} + V\frac{\partial A}{\partial x}\right)\phi(x,y)dxdy + \int_\Omega \rho\nu\left(\frac{\partial A}{\partial x}\cdot\frac{\partial V}{\partial x} + \frac{\partial A}{\partial y}\cdot\frac{\partial V}{\partial y}\right)dxdy = 0 \tag{16}$$

where ∅ is required to vanish wherever essential boundary conditions apply to $\overline{A}$ on the boundary. Substituting the finite element approximation of $\overline{A}$ in (16), yields the time dependent matrix equation

$$[M]\frac{d}{dt}[A] + [N][A] = 0 \tag{17}$$

which was solved by a time stepping technique.

Csendes and Chari [28] describe a general variational formulation for the diffusion problem, including the rotational term. This generalization of the eddy current problem is well suited for analyzing the asynchronous performance of rotating electrical machines.

In this method, the partial differential equation and its functional formulation are given by

$$\nu\nabla^2 A = \frac{1}{\rho}\ \frac{\partial A}{\partial t} + V\ x\ B - J_s \tag{18}$$

$$\mathscr{F} = \int_R \left|\nabla A\right|^2\ dR + \frac{\mu\omega}{\rho}\int_R \left(y\ \frac{\partial A}{\mu x} - x\ \frac{\partial A}{\partial y}\right)dR + j\ \frac{\mu\omega}{\rho}\int_R A^2 dR - 2\int_R A\cdot J dR \tag{19}$$

The above technique was applied to a simplified generator cross-section discretized by first order triangular elements, and a magnitude plot of the flux was presented.

Konrad and Chari [26] described a skin-effect problem by an integrodifferential formulation of the form

$$\nu\nabla^2\overline{A}\ - j\omega\ \sigma\overline{A} = -\ \frac{(I + j\omega\sigma\int\int\overline{A}\cdot\overline{ds})}{\int\int ds} \tag{20}$$

The solution obtained by the weighted residual procedure for a transmission line bundle conductor arrangement is illustrated by the magnitude plot of A as shown in Figure 7.

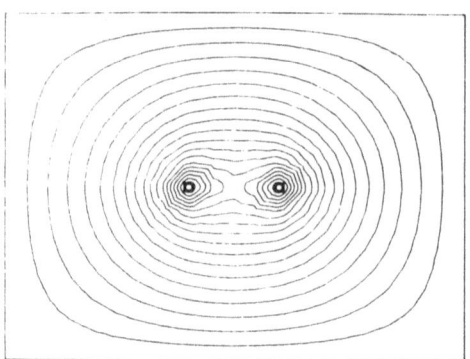

Figure 7. Flux distribution around an asymmetric five-wire
bundle conductor arrangement. The ac/dc loss
ratio for this system is 1.146 when the total current is 1000A at 60 Hz. The conductors do not
carry equal currents.

4. THREE-DIMENSIONAL SCALAR POTENTIAL SOLUTION

A three-dimensional scalar potential formulation of a disc type permanent magnet motor using samarium-cobalt magnets was described by Chari and D'Angelo [13]. The differential equation and functional expression for the magnetostatic problem were obtained as follows.

$$\text{div}\,(\mu_r \mu_o \,\text{grad}\,\Phi) = -\,\mu_o \text{div}\,\overline{M}_o \tag{21}$$

$$\mathscr{F} = \int_n \mu_r \mu_o (\text{grad}\,\Phi)^2 d\Omega + 2\int_n \mu_o \Phi \text{div}\,\overline{M}_o \,d\Omega \tag{22}$$

The vector plot of flux-densities over one pole-pitch and the comparison of the numerical values obtained with test results are illustrated in Figures 8(a) and 8(b), respectively.

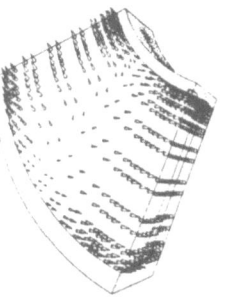

Figure 8(a).   Vector Plot of Flux Density
over One Pole-Pitch of
the Disc Machine

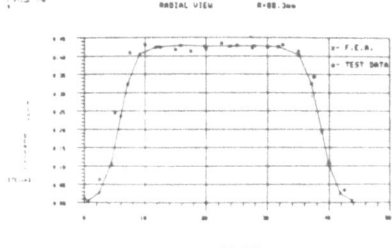

Figure 8(b).   The Flux-Density
Variation at the
Radial Line over the
Midplane of the Pole

5. CONCLUSIONS

In this paper, the finite element differential equation approach for modeling the electromagnetic fields in electrical machinery is presented and its application to practical engineering design problems is illustrated. Present day numerical methods by and large perform complementary functions, and it is emphasized that the designer and the analyst pay careful attention to the modeling aspects of the field problem and to the appropriate choice of one or the other of the solution techniques available.

13

REFERENCES

1. G.W. Carter, The Electromagnetic Field in its Engineering Aspects, Longmans Green & Co., London 1954.
2. O.V. Tozoni, Mathematical Models for the Evaluation of Electric and Magnetic Fields, Iliffe Books Ltd., London 1968.
3. E.F. Fuchs and E.A. Erdelyi, "Nonlinear Theory of Turboalternators," Parts I & II, IEEE Trans. Vol. PAS-92, No. 2, 1973, pp. 533-592.
4. J. Simkin and C.W. Trowbridge, "Application of Integral Equation Methods for the Numerical Solution of Magnetostatic and Eddy Current Problems," Rutherford Laboratory Report RL-76-041, U.K.
5. Z.J. Csendes, "The High Order Polynomial Finite Element Method in Electromagnetic Field Computation, Ch 7, Finite Elements in Electrical and Magnetic Field Problems, John Wiley & Sons., Ltd., U.K. 1980.
6. M.V.K. Chari and P. Silvester, "Analysis of Turbo-Alternator Magnetic Fields by Finite Elements," IEEE Trans, PAS-90, 1971, pp. 454-64.
7. P. Brandl, K. Reichert and W. Vogt, "Simulation of Turbogenerators on Steady State Load," Brown Boveri Review, 9, 1975.
8. S.H. Minnich, Z.J. Csendes, M.V.K. Chari, S.C. Tandon and J.F. Berkery, "Load Characteristics of Synchronous Generators by the Finite Element Method," F 80 206-3, IEEE PES, Winter Meeting, 1980.
9. Z.J. Csendes and A. Konrad, "Electrical Machine Modeling and Power System Constraints," Conference Proceedings on Electric Power Problems: The Mathematical Challenge, SIAM, Philadelphia, 1980.
10. A. DiMonaco, G. Giuseppetti, and G. Tontini, "Studio Di Campi Elettrici E Magnetici Stanzionari Con Il Metodo Degli Elementi Finiti Applicazione Ai Transformatori," L'Elettrotecnica, LXII, No. 7, pp. 585-598, 1975.
11. S.C. Tandon, "Finite Element Analysis of Induction Machines," IEEE Trans. Magnetics, Vol. MAG-18, No. 6 Nov 1982, pp. 1722-24.
12. M.V.K. Chari and P. Silvester, "Finite Element Analysis of Magnetically Saturated DC Machines," IEEE Paper 71 TP 3-PWR, Winter Meeting, February 1971.
13. M.V.K. Chari and J. D'Angelo, "Finite Element Analysis of Magneto-Mechanical Devices," Proceedings of the Fifth International Workshop on Rare Earth Cobalt Magnets and their Applications, Roanoke, Virginia, June 7-10, 1981, pp. 237-257.
14. T.G. Kincaid and M.V.K. Chari, "The Application of Finite Element Analysis Method to Eddy Current Nondestructive Evaluation," IEEE Trans. Vol. MAG-15, No. 6, pp. 1956-60, 1979.
15. M.A. Palmo, Z.J. Csendes and M.V.K. Chari, "Axisymmetric and Three Dimensional Electrostatic Field Solutions by the Finite Element Method," Elect. Machi. Electromech. Q., pp. 235-244, 1979.
16. M.V.K. Chari, "Finite Element Analysis of Nonlinear Magnetic Fields in Electric Machines," PhD Dissertation, McGill University, Montreal, Canada, 1970.
17. P. Silvester, H.S. Cabayan, and B.T. Browne, "Efficient Techniques for Finite Element Analysis of Electric Machines," IEEE Trans. Vol. PAS-92, No. 4, 1973.
18. E.H. Cuthill and J.M. McKee, "Reducing the Bandwidth of Sparse Symmetric Matrices," Proc. 24th Conference ACM, pp. 157-172, 1969.
19. K.J. Bathe and E.L. Wilson, Numerical Methods in Finite Element Analysis, Prentice Hall, Englewood Cliffs, N.J. 1976.
20. B.M. Irons, "A Frontal Solution Program for Finite Element Analysis," Int. Journal of Num. Methods in Engg., Vol. 2, 5-32, 1970.
21. M.R. Hestenes and E. Stiefel, "Method of Conjugate Gradients for Solving Linear Systems," Report No. 1659, NBS, Washington, D.C., 1952.
22. P. Silvester and C.R.S. Haslam, "Magnetotelluric Modelling by the Finite Element Method," Geophysical Prospecting, Vol. 20, pp. 872-891, 1972.
23. M.V.K. Chari, "Finite-Element Solution of the Eddy-Current Problem in Magnetic Structures," IEEE Trans., Vol. PAS-93, No. 1, 1973.
24. S.H. Minnich, J.W. Dougherty and D.K. Sharma, "Calculation of Generator Simulation-Model Constants using Finite-Element Analysis," ICEM, Budapest, Hungary, Sep. 1982.
25. J. Donea, S. Giuliani, and A. Philippe, "Finite Elements in the Solution of Electromagnetic Induction Problems," International Journal for Numerical Methods in Engineering, Vol. 8, 1974, pp. 359-367.
26. A. Konrad and M.V.K. Chari, "Power Loss and Forces Due to Skin Effect in Transmission Lines and Busbars," IEEE Canadian Communications & Energy Conference, Oct. 13-15, Montreal, 1982.
27. Z.J. Csendes and M.V.K. Chari, "General Finite Element Analysis of Rotating Electric Machines," International Conference on Numerical Methods in Electrical and Magnetic Field Problems, St. Margherita, Italy, 1976.
28. A. Foggia, J.C. Sabonnadiere, and P. Silvester, "Finite Element Solution of Saturated Travelling Magnetic Field Problems," IEEE Trans., Vol. PAS-94, No. 3, 1975.

A DETAILED LUMPED-PARAMETER METHOD FOR
THE STUDY OF ELECTRICAL MACHINES IN
TRANSIENT STABILITY STUDIES

L. Haydock and D.C. MacDonald

1. Trent Polytechnic, Nottingham
2. Imperial College, London

1. ABSTRACT

The turbine-generator flux decay test is modelled using a transient
analysis of the lumped-parameter equivalent-circuit model. The linked electric and magnetic circuits of the machine are represented by
components calculated from design data. Magnetic capacitors or magnetic admittances are defined to represent the magnetic flux paths in
the machine and the effects of eddy currents in the dampers and rotor
iron are represented by magnetic resistors. The transient analysis
of the flux decay for the magnetic circuit is performed in electrical
terms, using the electronic circuit analysis program Spice 2, and
used to calculate the machine terminal voltage during the quadrature-
axis flux decay test. Computed and test results on a 500MW set compare well. The method is simple and gives insight into machine transient behaviour as well as enabling machine parameters to be calculated. The identity of each component of the machine is preserved in
the model.

2. INTRODUCTION

The use of finite element methods for modelling the behaviour of electrical machines has found favour with many researchers and machine
designers, because of the possibility of very detailed studies of individual machines or proposed machine designs. Power systems analysts, however, show a marked reluctance to abandon lumped-parameter equivalent-circuit models, which is understandable for multi-machine studies, even though such models often fail to give accurate results and
the parameter values when obtained from different tests are often in
conflict, or produce simulations which do not even accurately reproduce the tests from which they were obtained.

A method of modelling the behaviour of electrical machines is proposed which stands between the full finite element field solution, with
its considerable demands on computing resources,and the supposedly
simple two-axis 'equivalent-circuit' models with their oversimplified
and coarsely lumped magnetic representation. Methods for the conver-

sion of magnetic circuits to equivalent electric forms suggested by Carpenter (1968) have been developed for the study of electrical mac hines, and in particular turbine-generators, by the authors and are de scribed in detail in Haydock (1978), (1980) and (1981) where it is shown that account may be taken of the complexities of the machine structure to produce topologically accurate models in which the iden tity of each component is preserved. Equivalent circuits of any com plexity may be produced in a routine fashion, once the basic princip les are grasped, and the conventional equivalent-circuit models of tl established two-axis theory, together with less conventional models such as those of Widger (1968) and Canay (1977), and others, may be de rived in a simpler way than was employed originally in most cases. Usually the simplification of the more detailed models requires gross approximations. The method also gives insight into the relationship between the electric and magnetic quantities in a way which is often obscured by more conventional methods of analysis and illustrates wh attempts to modify the conventional two-axis models by simply adding extra components in an ad hoc way, often with difficult justificatio is likely to fail or be misleading. A brief introduction to the met hods is presented here and the application to the study of the quadr ature axis flux decay test on a 500MW turbine-generator is described.

3. THE REPRESENTATION OF MAGNETIC AND ELECTRIC CIRCUITS

Many different analogue representations of magnetic circuits have be suggested. The most useful of them maintain the power dissipation o energy storage characteristics of the elements they represent. This is achieved here by taking the following analogy:

The rate of change of flux, usually with respect to time,

$(\frac{d}{dt} \phi)$ or $(p\phi)$ is taken as the magnetic flow quantity of flow function, and corresponds to the magnetic current i_m. Here $(p = \frac{d}{dt})$ or
may be used as the Heaviside or Laplace operator when operational
Calculus is employed. The magnetomotive force (F) is taken as the
potential function and corresponds to the magnetic voltage (v_m).

This leads to the representation of magnetic flux (q_m) in the magnetic circuit as analogues to electric charge (q) in the electric cir
cuit and the concept of a magnetic capacitor C_m which stores energy
in the form of a magnetic field between its terminals. The magnetic
capacitance is defined as the reciprocal of a pure (air-gap) magnetic
reluctance (S). Hence:

$$C_m \triangleq \frac{1}{S} \qquad (1)$$

The advantages over the more usual representation of magnetic reluctances as 'resistors' in magnetic circuits, and therefore energy dissipative instead of energy storing are obvious, and for pure reluctances the following magnetic circuit relationships which are immediately analogous to the corresponding electric circuit relationships, amongst others too numerous to state here, follow naturally:

$$i_m = \frac{d\varnothing}{dt} = p\varnothing = \frac{d}{dt} q_m \qquad (2)$$

$$i_m = C_m \frac{d}{dt} v_m = C_m \frac{dF}{dt} \qquad (3)$$

$$F = S\varnothing, \therefore F = \frac{1}{Cm} \int p\varnothing \cdot dt. = \frac{1}{Cm} \int pq_m \cdot dt.$$

$$\underline{OR} \qquad v_m = \frac{1}{Cm} \int i_m \cdot dt. \qquad (4)$$

By using the principle of duality the magnetic capacitors used as lumped parameters on the magnetic side of linked magnetic and electric circuits may be converted to magnetic-dual inductance parameters L_{mD} on the magnetic side or may be referred through the linkage and be represented by conventional electric inductance parameters L. More generally, where the magnetic circuit contains not simply pure (airgap) reluctances, but lossy reluctances, as in iron, the magnetic circuit lumped parameters are described more accurately as magnetic admittances (Y_m), which may be parallel or series combinations of magnetic capacitors and magnetic resistors, either of which may be non-constant value parameters in order to model non-linear effects such as saturation and losses.

Linkages and the half-transformer principle

If a magnetic and electric circuit are linked the linkage (N) is a reciprocal parameter and impedances, sources, or whole networks may be referred across the linkage either way to the electric or magnetic terminals. When this is done the linkage acts as a 'gyrator', potential and flow functions and sources becoming interchanged and series and parallel elements being transposed. Furthermore, impedance and admittance elements become multiplied or divided by N^2 and potential or flow functions are multiplied or divided by N, as appropriate. The magnetic and electric circuits may be separated by the 'tearing principle' described by Kron (1937), except rather than using connecting tensors for the simple cases considered here the half-transformer principle is employed. The half-transformer principle is so called because it describes the behaviour of a single electric-magnetic linkage, as described above, whereas its application twice between two electric circuits linked by a magnetic circuit is the familiar transformer principle. The familiar referral of impedances across a conventional transformer double linkage, though convenient for simple

17

cases, misses out a vital stage, the accurate representation of the
magnetic circuit, and in all but the most simple devices may produce
misleading models. It is this principle applied where it is inappropriate which is the fundamental weakness of conventional two-axis models. The half-transformer principle is illustrated by the referral
of an electric potential source and series impedance across a linkage
N into the linked magnetic circuit in figures 1 and 2.

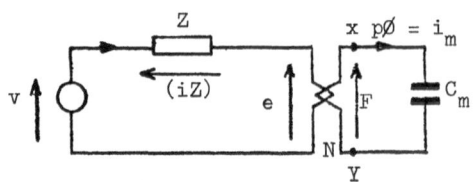

Figure 1. Linked electric and magnetic circuits

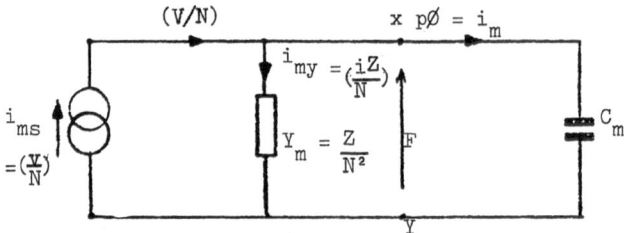

Figure 2. Electric circuit referred into the magnetic
circuit

The electric potential source (v) on the electric side of the linkage
N of figure 1 becomes a magnetic flow function or current source,

$$i_{ms} = \frac{v}{N} \qquad\qquad (5)$$

when referred across the linkage to give the wholly magnetic equivalent circuit of figure 2. The series electric impedance (Z) becomes the
parallel magnetic admittance,

$$Y_m = \frac{Z}{N^2} \qquad\qquad (6)$$

when referred across the half-transformer linkage. On the electric
side of the linkage of figure 1 evidently:

$$v = e + iZ \qquad\qquad (7)$$

$$\text{where} \quad e = Np\emptyset \qquad\qquad (8)$$

To refer the electric source to the magnetic side of the linkage

18

equation (7) is divided by N and the 'gyration' takes place so that:

$$\frac{v}{N} = \frac{iZ}{N} + \frac{Np\emptyset}{N} \qquad \underline{OR}$$

magnetic current = magnetic current in + magnetic current into
 source   magnetic admittance   magnetic capacitor

$$i_{ms} = i_{my} + p\emptyset \qquad\qquad (9)$$

This shows that the magnetic flow quantity into the magnetic circuit
on the right-hand side of the terminals xy in figure 2 is the same
as in figure 1. For the same magnetic potential $F = v_m$ across the
terminals xy in figures 1 and 2, then:

$$Y_m = \frac{i_{my}}{v_m} = \frac{i.Z}{N} \cdot \frac{1}{iN} = \frac{Z}{N^2}$$

$$\underline{OR} \qquad Y_m = \frac{Z}{N^2} \qquad \text{as stated in (6)}$$

These basic methods may be extended to deal with much more complicated electric and magnetic circuits and fully referred equivalent circuits of either wholly electric **or** wholly magnetic form may be derived,
depending upon the desired circuit form and the analysis to be performed. Although, wholly electric circuits are the most familiar, it
has been found that the 'direct magnetic circuit', i.e. that containing magnetic admittances and capacitors is the most immediately useful for the study of flux decay tests. The term 'direct magnetic
circuit' is used to distinguish this type of magnetic circuit from the
magnetic-dual form containing magnetic-dual impedances and inductances
described by Haydock (1980) which may be derived by taking the dual
of the direct magnetic circuit. The magnetic dual form has uses, especially when fully referred electric equivalent circuits are desired,
but the direct magnetic form is the type used in the following analysis of the quadrature axis flux decay test on a turbine-generator.

4. FLUX DECAY TESTS

Flux decay tests are performed on Turbine-generators at realistic flux
and current levels and can give information about the transient behaviour of any axis of the machine. The test involves the exciting of
the stator from the system and rotating the rotor synchronously to
align any rotor axis with the field. The stator excitation is then
suddenly removed and the decaying stator voltage induced by the decaying flux, which is supported by eddy currents in the iron and winding

19

circuits of the rotor, is monitored. The resulting waveform may be used to extract machine parameters.

The quadrature axis flux decay test

The first condition that has been studied is the quadrature axis flux decay test where the main field axis is aligned with the quadrature axis of the rotor, before the stator excitation is removed. Figure 3 represents these conditions and shows a half pole pitch view of a 500 MW Turbine-generator cross-section (not to scale). Superimposed on the machine cross-section is a simplified grid of possible magnetic flow paths, shown dotted, connecting up magnetic circuit components of the types described previously. The airgap reluctances are represented by pure magnetic capacitors 1. Rotor tooth-top leakage paths are represented by magnetic capacitors 2, and stator leakage by magnetic capacitors such as 11, an example only is shown for clarity.

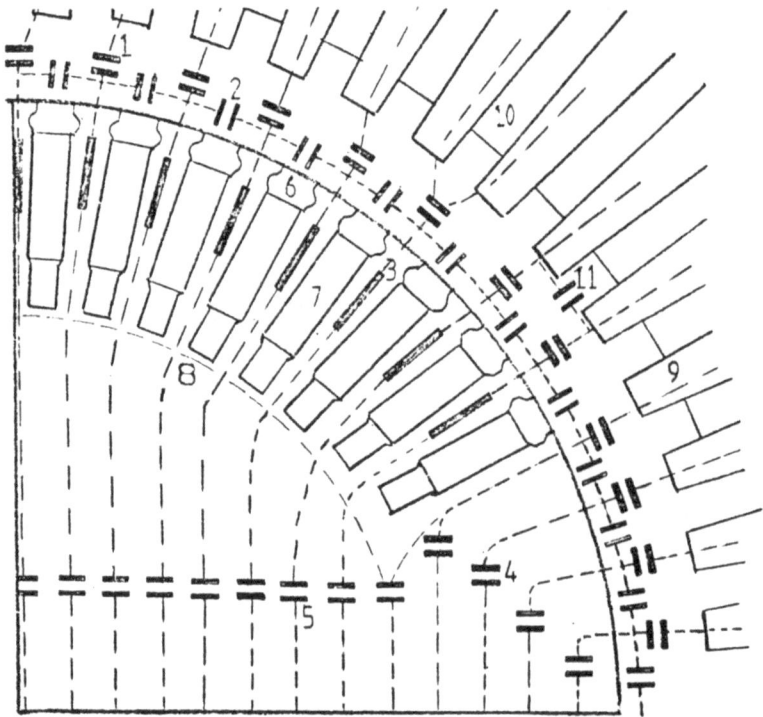

Figure 3. 500MW Turbo-generator cross-section with superimposed magnetic circuit

20

The reluctances of the lumped reluctance paths of the rotor iron are
represented by magnetic admittances such as those shown as shaded box-
es for tooth admittances 3, if a lossy or non-linear representation is
required. If a linear loss-free representation of iron may be toler-
ated the iron reluctance paths may be represented by pure magnetic
capacitors such as those for the pole region 4 or the rotor body region 5. There will be other flux paths, such as 8, with associated
magnetic components which are not shown for clarity. The values of
the magnetic components may be calculated by established design methods which are tedious, but straightforward. For example a magnetic
capacitor for a portion of the magnetic path:

$$C_m = \frac{1}{S} = \frac{\mu_0 \mu_r}{l} \, a \qquad\qquad (10)$$

The units are 'Magnetic Farads' or Webers per Ampere-turn.

Where μ_0 is the magnetic permeability of free space,
    μ_r is the relative permeability of the material, and
    a is the area and l the length of the flux path.

The choice of lumping of assumed flux paths, i.e. the discretization
of the machine parts and the choice of magnetic parameters such as
the relative permeability of iron paths require considerable engineering judgement and design experience. Here knowledge gained from finite element studies can be invaluable.

If the magnetic and electric circuits such as those formed by the damper Wedges 6, rotor iron, and stator winding 10 are torn from the magnetic circuit, then a linked magnetic and electric circuit model of
the type previously illustrated in figures 1 and 2, except with many
more components, results and the physical representation of the rotor
body and stator may be dispensed with, though the identity of each
circuit component and its relationship to the machine topology is preserved.

5. NUMERICAL RESULTS

Such equivalent circuit models have been analysed by using the electronic circuit analysis package Spice 2G.0, Vladimirescu (1980), on
the D.E.C. 20 Computer at Trent Polytechnic. The magnetic circuit is
initialized with potential and flow functions related to the pre-tra-
nsient conditions. The conditions in the magnetic circuit are analysed in electrical network terms and the fundamental of the airgap flux
is used to calculate the stator winding voltage at each point in the
decay. The results of studies using a model with 50 magnetic components and less than 50 magnetic nodes to simulate the quadrature axis
flux decay test on a 500MW Turbine-generator are shown compared with
test results in the graphs of figure 4. The C.P.U. time to simulate

21

the test was 6 seconds, with up to 200 points per plot, therefore, a
fine discrimination and smooth plot was possible, as was the detailed
examination of individual parts of the plot, such as the initial stages of the decay. The internal plotting routines, of Spice 2, can
give plots direct on the line printer or v.d.u. This was useful for
preliminary checks of the results.

The three-phase stator winding voltage before the breaker was opened
was 21.7kV (line) in the test modelled. Turner (1981) has shown that
the three-phase stator winding of a generator may be replaced by a
two-phase sinusoidally distributed winding which is insensitive to al
harmonics of the airgap flux waveform except the fundamental and that
the number of turns associated with a node is give by:

$$N_{di} = \frac{6}{\pi} N_{eff} \frac{\Delta i \; Sin \; \theta i}{\sum\limits_{j=1}^{n} | \Delta j \; Sin\theta_j |} \qquad (11)$$

and,

$$N_{qi} = \frac{-6}{\pi} N_{eff} \frac{\Delta i \; Cos \; \theta i}{\sum\limits_{j=1}^{n} | \Delta j \; Cos\theta_j |} \qquad (12)$$

where N_{eff} = effective series turns per pole pair per phase

The subscript d refers to a direct axis winding and q to a quadrature
axis winding.

Δi = area of conductor associated with the i th node
θi = angle from the direct axis of the i th node.

Similarly, the fundamental flux $\emptyset_f$ to establish the excitation conditions for the quadrature axis flux decay test may be established:

$$\emptyset_f = \frac{\sum\limits_{i=1}^{n} \emptyset_i \; Sin \; \theta i \; \delta\theta i}{\sum\limits_{i=1}^{n} | \delta\theta i \; Sin \; \theta i |} \qquad (13)$$

where θi is the angle from the axis of interest
$\delta\theta i$ is the increase in angle
$\emptyset_i$ is the flux at node i

The airgap voltage V_G at the start of the transient may be calculated
by subtracting the stator leakage volt-drop $(I.X_1)$ from the stator
voltage before the stator is de-energised (V_L); hence:

22

$$V_G = V_L - \sqrt{3}\ IX_1 \tag{14}$$

Also, for a given winding $V_G = K\ \emptyset_f$ (15)

where K is a constant depending upon machine design details.

These relationships are used to calculate the stator voltage at each time step in the transient decay.

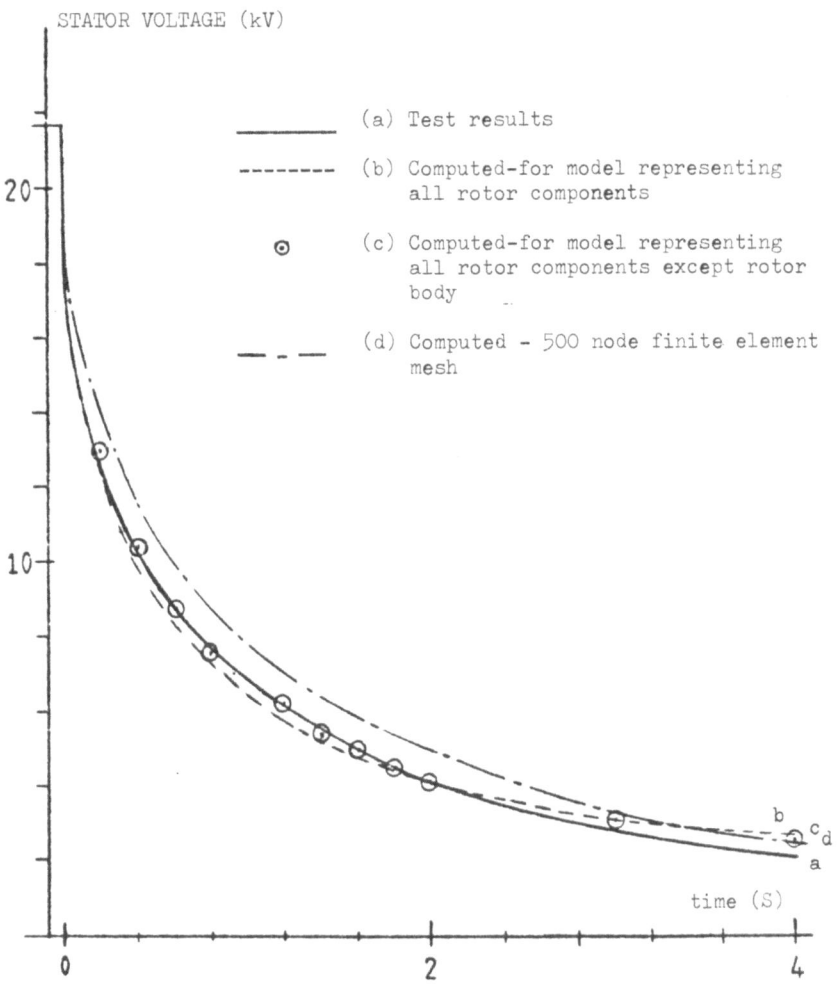

STATOR VOLTAGE (kV)

(a) Test results

(b) Computed-for model representing all rotor components

(c) Computed-for model representing all rotor components except rotor body

(d) Computed - 500 node finite element mesh

time (S)

Figure 4. Q-Axis flux decay test: test and calculated results

Discussion of the results

The graphs of figure 4 show that the calculated terminal voltages agree with the measured terminal voltages very well in all the cases illustrated. The agreement is least good at the longer times. The best fit of the results illustrated is that for a model containing components to represent the rotor teeth, wedges and pole iron, but not the rotor body, the points (c) in figure 4. One advantage of the method is that it allows the modelling of the behaviour with only certain components present, e.g. only the rotor wedges. These studies, and the results depicted in figure 4, show that in the case of the quadrature axis flux decay test the iron pole eddy current paths dominate the decay, especially in the long term, and this is thought to be the cause of the previously unexplained long time constant observed by Shackshaft (1977). The curve for a model with all rotor components represented, including the wedges, teeth, pole and rotor body, curve (b) in figure 4, is a perfect initial fit to the test results up to 0.2 seconds, it then decays slightly too fast up to a time of 2 seconds, after which it decays more slowly than the curve for the test results, curve (a), which is a mean curve plotted through the test points.

In curve (b) and points (c) of figure 4, the magnetic circuit contained only constant parameter elements. Saturation was allowed for only by the choice of different values of the relative permeability, μ_r, at different points in the model. A value of $\mu_r = 60$ was used for the pole iron in both cases, whereas a μ_r of 100 was used for the rotor body in (b). Another simulation was performed by making $\mu_r = 150$ for the rotor body and gave a slightly worse initial fit than (b), but a slightly better long time fit. Evidently, a near perfect fit to any test results could be achieved by having either circuit elements with non-constant parameters, or grading the parameters at different points in the circuit or at different times in the decay. Clearly μ_r does vary in the rotor as the decay proceeds and the magnetic saturation in the rotor iron changes. However, for the test simulated here the agreement between the calculated and test results is considered to be within the measurement tolerances, and, therefore, the extra complication of a non-constant parameter model is not warranted.

For comparison purposes curve (d) is reproduced from Turner (1981), figure 6.15, and shows the calculated results from a 500 node finite element model for the same machine. The curve (d) is a worse fit on a point by point basis than the lumped parameter model results, especially in the early and middle portion, but is more parallel to the test results, especially in the long term. It was at first thought that the error in the finite element results was due to too coarse a mesh, but a 997 node mesh with a pole surface discretization of 2mm accounted for only about one third of the error. Based on this knowledge, and other design considerations, the pole was modelled in graded depths of 0.3 mm upwards in the magnetic circuit model. It is not intended to imply that the lumped parameter equivalent circuit model is superior to the finite element method. Both methods have their uses and are complementary.

Machine parameters

The results from the model simulations may be used to produce machine parameters including time constants as in the C.E.G.B. Report (1976). However, rather than reproduce the three approximate time constants derived there, it is noted that the rate of decay varies continuously and there are a great many, if not an infinite number, of time constants present, though dominant time constants may be extracted, if required.

6. CONCLUSIONS

The method proposed has produced encouraging results when applied to the quadrature axis flux decay tests. It is at present being applied to other tests and other machines. The method has the advantage of being simple and the demands on computing facilities are extremely modest. Models of varying complexity may be used for different purposes, and the method may be used to investigate the effects of different parts of the machine on transient behaviour, because the identity of each component is preserved in the topologically accurate equivalent circuit models.

7. ACKNOWLEDGEMENTS

Thanks are due to A.B.J. Reece and Dr. P.J. Turner of G.E.C. Limited, Stafford, for providing the details of the 500MW generator and for other help and advice; Mr. W. Fairney and C.E.G.B. for supporting the work and permission to publish the test results; the S.E.R.C. for support in the initial stages of the work; Prof. E.M. Freeman, Dr. C.J. Carpenter and others at Imperial College for help and advice; Dr. R.W. Burns, the Computer Services Department and Trent Polytechnic for support and the use of computing facilities.

8. REFERENCES

(1)   Carpenter, C.J. 'Magnetic equivalent circuits', Proc. I.E.E., Vol. 115, No. 10, pp 1303-1511, (October, 1968).

(2)   Canay, I.M., 'Extended synchronous machine model for calculation of transient processes and stability', Electrical machines and electromechanics, Vol. 1, pp 137-150, (1977).

(3)   C.E.G.B. (Barber, M.D. et al), 'Report of generator parameter tests at Eggborough Power Station', Report PL-ST/11/76 (1976).

(4)   Haydock, L. (Haddock, L.), 'Alternator parameters', D.I.C. Thesis Imperial College, London, (1978).

(5)   Haydock, L. (Haddock, L.), 'Magnetic circuits for the operational impedances of synchronous generators from basic principles', Power Systems Report No. 103, Imperial College, London, (Machines and Power Systems Section), (April 1980).

(6)   Haydock, L. (Haddock, L.), 'Equivalent circuits for synchronous machines by direct and magnetic dual methods of modelling', Easter Course: synchronous machines in power systems, Imperial College, London, (March 1981).

(7)   Kron, G., 'The application of tensors to the analysis of rotating electrical machinery', General Electric Review, a serial (April 1935 to December 1937).

(8)   Shackshaft, G. and Poray, A.T., 'Implementation of new approach to the determination of synchronous-machine parameters from tests', Proc. I.E.E., Vol. 124, No. 12, pp 1170-1178 (December 1977).

(9)   Vladimirescu, A. et al, 'Spice Version 2G.0. User's Guide', Department of Electrical Engineering and Computer Sciences, University of California, Berkeley, Ca., (September 1980).

(10) Turner, P.J., 'Finite element electromagnetic analysis of turbine-generator performance', Ph.D Thesis, Imperial College, London, (November 1981).

(11) Widger, G.F.T. and Adkins, B., 'Starting performance of synchronous motors with solid salient poles', Proc. I.E.E., Vol. 115, No. 10, pp 1471-1484, (October 1968).

THE ELECTROMAGNETIC FIELD OF AN INSULATED CONDUCTOR IN A SLOT

S G Mudge

The Polytechnic, Wolverhampton

1. INTRODUCTION

The model considered is that of a slot infinite in length in the axial direction and of rectangular cross-section. It is surrounded on three sides by iron core and the remaining side is open. The insulated conductor, also infinite in length and of rectangular cross-section, is entirely within the slot. A current, varying sinusoidally with time, flows in the axial direction. Such a configuration occurs in practice in the design of slot wound electrical machines and much of the theoretical work to date on this subject has assumed that the effect of the insulation space is negligible. The purpose of the study described here is to examine the validity of this assumption.

Using Maxwell's equations, the field equations to be satisfied in both the conductor and the insulator are derived together with appropriate boundary conditions. A solution, valid throughout the whole region of the slot, is developed in terms of the vector potential. This solution is in the form of a two dimensional truncated Fourier series. The Fourier coefficients are determined from a set of linear complex equations for which an iterative method of solution is described. The rate of convergence of the Fourier series is examined and an algorithm for efficient summation of such series is described.

The impedance of the conductor is found by integration of the Poynting vector over the surface of the slot. The resistance and inductance at zero frequency are given in closed form and so demonstrate one advantage of this model. Most techniques used for the solution of the type of problem considered here evaluate the inductance for a sufficiently small value of the frequency.

2. LIST OF SYMBOLS

$\underline{E}$	electric intensity	(volt/metre)
$\underline{B}$	magnetic flux density	(weber/square metre)
$\underline{D}$	electric displacement density	(coulomb/square metre)
$\underline{H}$	magnetic intensity	(ampere-turn/metre)
$\underline{J}$	current density	(ampere/square metre)
ρ	volume charge density	(coulomb/cubic metre)
ε_0	permittivity of free space	(8.854×10^{-12} farad/metre)

μ_o	permeability of free space	($4\pi\times10^{-7}$ henry/metre)
μ_r	relative permeability	
ε_r	relative permittivity	
σ	conductivity	(siemens/metre)
V	scalar potential	(volt)
$\underline{A}$	vector potential	(weber/metre)
α	$\sqrt{\mu_r\,\mu_o\,\sigma\,\omega}$	
ω	angular frequency	(radians/second)
I	current	(ampere)
W	magnetic energy	(joules)
L	leakage inductance	(henry)
R	effective resistance	(ohm)
X	inductive reactance	(ohm)
*	denotes complex function	
$\sim$	denotes complex conjugate	
∇^2	Laplacian operator	
$\sum_{n=0}^{'}$	denotes that a factor $\frac{1}{2}$ is to be taken when n=0.	

3. DERIVATION OF THE MATHEMATICAL MODEL

The slot configuration is as shown in Figure 1 and the differential equations to be satisfied in the conductor and the insulator are derived from Maxwell's equations.

$\frac{\partial D}{\partial t}$ is neglected; the usual assumption for the range of frequencies to be considered. Then defining the vector potential $\underline{A}$ by

$$\underline{B} = \text{curl } \underline{A} \text{ where div } \underline{A} = 0, \quad \nabla^2\underline{A} = -\mu_r\,\mu_o\,\underline{J} \qquad (1)$$

since curl $\underline{B} = \mu_r\,\mu_o\,\underline{J}$. Since curl $\underline{E} + \frac{\partial \underline{B}}{\partial t} = \underline{0}$, there is a scalar function V such that, to within a constant,

$$\underline{E} + \frac{\partial \underline{A}}{\partial t} = -\text{ grad } V \qquad (2)$$

Vector potential in the insulator
Here the current density is everwhere zero, i.e. $\underline{J} = \underline{0}$ and Equation (1) becomes

$$\nabla^2\underline{A} = \underline{0} \qquad (3)$$

Assuming sinusoidal variation with time t,

$$\underline{A} = R\ell(e^{i\omega t}\,\underline{A}*), \quad i = \sqrt{-1} \qquad (4)$$

28

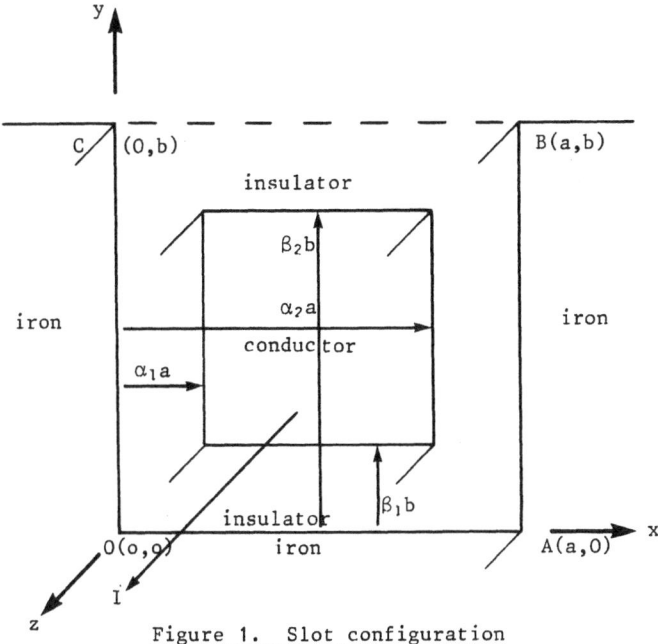

Figure 1. Slot configuration

where $\underline{A}^*$ is a complex vector function of position and $R\ell(z)$ denotes the real part of z. With the configuration of Figure 1.,

$$\underline{A}^* = (0, 0, A_z^*)\tag{5}$$

and so, combining Equations (3), (4) and (5), the differential equation to be satisfied in the insulator is

$$\frac{\partial^2 A_z^*}{\partial x^2} + \frac{\partial^2 A_z^*}{\partial y^2} = 0\tag{6}$$

Vector potential in the conductor
Equation (2) may be written as $\underline{E} = \underline{E}_1 + \underline{E}_2$ where $\underline{E}_1 = -\text{grad } V$ and $\underline{E}_2 = -\dfrac{\partial \underline{A}}{\partial t}$.

From Ohm's law, $\underline{J} = \sigma\underline{E} = \sigma(\underline{E}_1 + \underline{E}_2)$. Defining $\underline{J}_1 = \sigma\underline{E}_1 = -\sigma\text{grad } V$ and $\underline{J}_2 = \sigma\underline{E}_2 = -\sigma\dfrac{\partial \underline{A}}{\partial t}$,

Equation (1) becomes $\nabla^2 \underline{A} = -\mu_r \mu_o \left(\underline{J}_1 - \sigma\dfrac{\partial \underline{A}}{\partial t}\right)$

Combining this with Equations (4) and (5), the differential equation to be satisfied in the conductor is

$$\frac{\partial^2 A_z^*}{\partial x^2} + \frac{\partial^2 A_z^*}{\partial y^2} - i\alpha^2 A_z^* = -\mu_r \mu_o J_{1z}^* \qquad (7)$$

since $\underline{J}_1 = R\ell(e^{i\omega t} \underline{J}_1^*)$ and $\underline{J}_1^* = (0, 0, J_{1z}^*)$

Boundary conditions

The boundary conditions across a surface dividing two media are
 (i) the normal component of $\underline{B}$ is continuous and
 (ii) the tangential components of $\underline{H}$ are continuous across a change in medium when the conductivity is bounded. In practice the permeability of the insulator is the same as that of the conductor so that across the insulator-conductor boundaries, the tangential and normal components of both $\underline{B}$ and $\underline{H}$ are continuous. Thus throughout the interior of the slot $\frac{\partial A_z^*}{\partial x}$ and $\frac{\partial A_z^*}{\partial y}$ are continuous. This implies that A_z^* must be continuous also. The iron is assumed to have infinite permeability so that at the iron-insulator boundaries the tangential components of $\underline{H}$ must be zero. This gives rise to the following boundary conditions:

$$\frac{\partial A_z^*}{\partial x} = 0 \text{ on OC, AB} \qquad (8)$$

$$\frac{\partial A_z^*}{\partial y} = 0 \text{ on OA} \qquad (9)$$

It is assumed that along BC, A_z^* is constant, i.e. $A_z^* = \Omega$ on BC   (10) where Ω is a complex constant.

Sufficient equations are now given for A_z^* to be determined throughout the slot.

4. METHOD OF SOLUTION

Writing $B_z^* = A_z^* - \Omega$, the boundary conditions on B_z^* become

 (i) $\frac{\partial B_z^*}{\partial x} = 0$ for $x = 0, a$; $0 \leqslant y \leqslant b$

 (ii) $\frac{\partial B_z^*}{\partial y} = 0$ for $y = 0$, $0 \leqslant x \leqslant a$

 (iii) $B_z^* = 0$ for $y = b$, $0 \leqslant x \leqslant a$

The differential equations for B_z^* are

$$\frac{\partial^2 B_z^*}{\partial x^2} + \frac{\partial^2 B_z^*}{\partial y^2} - i\alpha^2 B_z^* = -\mu_r \mu_o J_{1z}^* + i\alpha^2 \Omega \text{ in the conductor, and} \qquad (11)$$

$$\frac{\partial^2 B_z^*}{\partial x^2} + \frac{\partial^2 B_z^*}{\partial y^2} = 0 \text{ in the insulator.} \qquad (12)$$

Using a double Fourier series, in order to satisfy the boundary conditions B_z^* must be of the form

$$B_z^* = \sum_{m=o}^{\infty} {}' \sum_{k=o}^{\infty} C_{mk} \cos \frac{m\pi x}{a} \cos (k+\tfrac{1}{2}) \frac{\pi y}{b} \tag{13}$$

where each C_{mk} is a complex constant. If this is to satisfy the differential equation

$$\sum_{m=o}^{\infty} {}' \sum_{k=o}^{\infty} C_{mk} \left\{ \frac{m^2 \pi^2}{a^2} + (k+\tfrac{1}{2})^2 \frac{\pi^2}{b^2} + i\alpha^2 \right\} \cos \frac{m\pi x}{a} \cos (k+\tfrac{1}{2}) \frac{\pi y}{b}$$

$$= \begin{cases} (\mu_r \mu_o J_{1z}^* - i\alpha^2 \Omega) & \text{in the conductor} \\[2mm] i\alpha^2 \sum_{p=o}^{\infty} {}' \sum_{q=o}^{\infty} C_{pq} \cos \frac{p\pi x}{a} \cos (q+\tfrac{1}{2}) \frac{\pi y}{b} & \text{in the insulator} \end{cases}$$

Equations for the coefficients C_{mk} are obtained by multiplying this expression by $\cos \frac{m\pi x}{a} \cos (k+\tfrac{1}{2}) \frac{\pi y}{b}$ and integrating over the whole area of the slot.

Writing $D_{mk} = C_{mk} \dfrac{\pi^4}{4b^2 (\mu_r \mu_o J_{1z}^* - i\alpha^2 \Omega)}$ the equations relating the coefficients D_{mk} are

$$PP(m,k)D_{mk} + i \left(\frac{\alpha b}{\pi} \right)^2 \sum_{p=o}^{\infty} {}' \sum_{q=o}^{\infty} AA(p,m)BB(q,k)D_{pq} = \alpha(m)\beta(k)$$

$$m = 0,1,2 \ldots, \quad k = 0,1,2 \ldots \tag{14}$$

where $PP(m,k) = \left(\dfrac{b}{a} \right)^2 m^2 + (k+\tfrac{1}{2})^2$,

$$AA(p,m) = AA(m,p) = \frac{1}{(p+m)\pi} \{ \sin \alpha_2 (p+m)\pi - \sin \alpha_1 (p+m)\pi \}$$

$$+ \frac{1}{(p-m)\pi} \{ \sin \alpha_2 (p-m)\pi - \sin \alpha_1 (p-m)\pi \} \ (p \neq m)$$

$$AA(p,p) = \frac{1}{2p\pi} \{ \sin \alpha_2 2p\pi - \sin \alpha_1 2p\pi \} + (\alpha_2 - \alpha_1) \ (p \neq 0)$$

$$AA(0,0) = 2(\alpha_2 - \alpha_1)$$

$$BB(q,k) = BB(k,q) = \frac{1}{(q-k)\pi} \{ \sin \beta_2 (q-k)\pi - \sin \beta_1 (q-k)\pi \}$$

$$+ \frac{1}{(q+k+1)\pi} \{ \sin \beta_2 (q+k+1)\pi - \sin \beta_1 (q+k+1)\pi \} (q \neq k)$$

$$BB(k,k) = \frac{1}{(2k+1)\pi} \{ \sin \beta_2 (2k+1)\pi - \sin \beta_1 (2k+1)\pi \} + (\beta_2 - \beta_1)$$

$$\alpha(m) = \frac{\sin(\alpha_2 m\pi) - \sin(\alpha_1 m\pi)}{m} \ (m \neq 0), \qquad \alpha(0) = (\alpha_2 - \alpha_1)\pi$$

$$\beta(k) = \frac{\sin(\beta_2 (k+\tfrac{1}{2})\pi) - \sin(\beta_1 (k+\tfrac{1}{2})\pi)}{(k+\tfrac{1}{2})}$$

Since the terms D_{mk} are of the same order of magnitude along diagonals, the series is summed by diagonals, truncating at the M'th diagonal.

Writing $D_{mk} = R\ell(D_{mk}) + iIm(D_{mk})$, Equation (14) can be written

31

$$PP(m,k) \ R\ell(D_{mk}) - (\frac{\alpha b}{\pi})^2 \sum_{p=o}^{M} {}' \sum_{q=o}^{M-p} AA(p,m) \ BB(q,k) \ Im(D_{pq})$$
$$= \alpha(m) \ \beta(k) \qquad (15)$$
$$PP(m,k) \ Im(D_{mk}) + (\frac{\alpha b}{\pi})^2 \sum_{p=o}^{M} {}' \sum_{q=o}^{M-p} AA(p,m) \ BB(q,k) \ R\ell(D_{pq}) = 0$$

for m = 0,1,2 ... M, k = 0,1, ...(M-m)

An iterative procedure for solving a coupled set of equations of this type is given in Appendix B. Having calculated the coefficients D_{mk}, A_z^* is given by

$$A_z^* = \Omega + \frac{4b^2}{\pi^4} (\mu_r\mu_o J_{1z}^* - i\alpha^2\Omega) \sum_{m=o}^{M} {}' \sum_{k=o}^{(M-m)} D_{mk} \cos\frac{m\pi x}{a} \cos(k+\tfrac{1}{2})\frac{\pi y}{b} \qquad (16)$$

The Fourier series is summed using the algorithms described in Appendix A.

In order to have some measure of the degree of accuracy of the solution for a given M, the total axial current in the conductor was evaluated by two different methods. The answers should, of course, be identical but differ due to the truncation error. The magnitude of the difference is a measure of this error.

If I_z is the total current flowing in the conductor then $I_z = R\ell(I_z^* e^{i\omega t})$

and (i) $I_z^* = \int_{x=\alpha_1 a}^{\alpha_2 a} \int_{y=\beta_1 b}^{\beta_2 b} J_z^* \ dx \ dy = \int_{\alpha_1 a}^{\alpha_2 a} \int_{\beta_1 b}^{\beta_2 b} (J_{1z}^* - \sigma i\omega A_z^*) dx \ dy$

With A_z^* given by Equation (16),

$$\mu_r\mu_o \ I_z^* = (\mu_r\mu_o J_{1z}^* - i\alpha^2\Omega)ab \ \{(\alpha_2 - \alpha_1)(\beta_2 - \beta_1) - 4i \ (\frac{\alpha b}{\pi^3})^2 \sum_{m=o}^{M} {}' \sum_{k=o}^{M-m} D_{mk}\alpha(m)\beta(k)\} \qquad (17)$$

Also (ii) $\mu_r\mu_o \ I_z^* = - \int_0^a (\frac{\partial A_z^*}{\partial y})_{y=b} \ dx$. Using Equation (16) this becomes

$$\mu_r\mu_o I_z^* = \frac{2ab}{\pi^3} (\mu_r\mu_o J_{1z}^* - i\alpha^2\Omega) \sum_{k=o}^{M} D_{ok}(k+\tfrac{1}{2})(-1)^k \qquad (18)$$

Examination of these two expressions for I_z^* shows that the series in Equation (17) has a more rapid rate of convergence than that in Equation (18).

M	$R\ell(kI_z^*)$ using Equation (17)	$R\ell(kI_z^*)$ using Equation (18)	$-Im(kI_z^*)$ using Equation (17)	$-Im(kI_z^*)$ using Equation (18)
8*	0.1553	0.1585	0.2174	0.2177
12*	0.1553	0.1543	0.2174	0.2173
8†	0.0504	0.0531	0.0901	0.0907
12†	0.0503	0.0496	0.0902	0.0898

Table 1. Convergence of series

(* $\alpha b=3$, † $\alpha b=6$, $k = \dfrac{\mu_r \mu_o}{(\mu_r \mu_o J_{1z}^* - i\alpha^2 \Omega)ab}$)

The data used to obtain this table is $\alpha_1 = 0.2$, $\alpha_2 = 1$, $\beta_1 = 0.1$, $\beta_2 = 0.9$, $\dfrac{b}{a} = 3$, corresponding to the conductor being symmetrically placed in a slot of total width 2a.

Table 1 shows reasonable agreement between the values of I_z^* obtained by Equations (17) and (18) and this agreement is improved as more terms of the series are included. Also the series (17) has converged to within 0.2% for the range of frequencies to be considered and so the expression for I_z^* given by Equation (17) will be used in the subsequent impedance calculations.

5. EVALUATION OF COMPLEX IMPEDANCE

The complex Poynting vector is defined to be $\frac{1}{2} \underline{E}^* \times \underline{\tilde{H}}^*$

where $\underline{E} = R\ell(\underline{E}^* e^{i\omega t})$, $\underline{H} = R\ell(\underline{H}^* e^{i\omega t})$ and $\tilde{z}$ denotes the complex conjugate of z. $\underline{E}^*, \underline{H}^*$ are complex, vector functions of position only.

If S is a closed surface bounding a volume V and $d\underline{S}$ is in the direction of the outward normal to S,

$$\iint_S \frac{1}{2} (\underline{E}^* \times \underline{\tilde{H}}^*).d\underline{S} = \frac{1}{2} \iiint_V \text{div}(\underline{E}^* \times \underline{\tilde{H}}^*)dV$$

$$= \frac{1}{2} \iiint_V (\underline{\tilde{H}}^*.\text{curl }\underline{E}^* - \underline{E}^*.\text{curl }\underline{\tilde{H}}^*)dV$$

Now curl $\underline{E}^* = -i\omega \underline{B}^*$ and curl $\underline{\tilde{H}}^* = \underline{\tilde{J}}^*$.

$$\therefore \iint_S \frac{1}{2}(\underline{E}^* \times \underline{\tilde{H}}^*).d\underline{S} = -\frac{1}{2} i\omega \iiint_V \frac{\underline{\tilde{B}}^*.\underline{B}^*}{\mu_r \mu_o} dV - \frac{1}{2} \iiint_V \frac{\underline{J}^*.\underline{\tilde{J}}^*}{\sigma} dV$$

Now $\frac{1}{2} \iiint_V \frac{\underline{J}^*.\underline{\tilde{J}}^*}{\sigma} dV = \frac{1}{2} R I_z^* . \tilde{I}_z^*$ and $\frac{1}{2} \iiint_V \frac{\underline{\tilde{B}}^*.\underline{B}^*}{\mu_r \mu_o} dV = \frac{1}{2} L I_z^* . \tilde{I}_z^*$

$$\therefore \iint_S \frac{1}{2} (\underline{E}^* \times \underline{\tilde{H}}^*).d\underline{S} = -\frac{1}{2} I_z^*.\tilde{I}_z^*(R+iX) \text{ where } X = \omega L \qquad (19)$$

Now $\underline{E}^* = (0,0, \frac{1}{\sigma}J_{1z}^* - i\omega A_z^*)$ and $\underline{\tilde{H}}^* = \frac{1}{\mu_r \mu_o} (\frac{\partial}{\partial y} \tilde{A}_z^*, -\frac{\partial}{\partial x} \tilde{A}_z^*, 0)$ so that

taking account of the boundary conditions round the slot, the only contribution to the integral (19) arises from the side BC where $A_z^* = \Omega$.

Hence $-I_z^* \tilde{I}_z^*(R+iX) = \frac{1}{\mu_r \mu_o} \left(\dfrac{J_{1z}^*}{\sigma} - i\omega\Omega \right) \int_0^a \left(\dfrac{\partial \tilde{A}_z^*}{\partial y} \right)_{y=b} dx = -\tilde{I}_z^* \left(\dfrac{J_{1z}^*}{\sigma} - i\omega\Omega \right)$

If I_z^* is given by Equation (17),

33

$$R+iX = \cfrac{1}{\sigma ab \left\{ (\alpha_2-\alpha_1)(\beta_2-\beta_1)-4i \left(\cfrac{\alpha b}{\pi^3} \right)^2 \sum_{m=o}^{M} \sum_{k=o}^{M-m} D_{mk} \; \alpha(m)\beta(k) \right\}} \tag{20}$$

The resistance at zero frequency is $R_o = \cfrac{1}{\sigma ab(\alpha_2-\alpha_1)(\beta_2-\beta_1)}$ and thus the ratios $\cfrac{R}{R_o}, \cfrac{X}{R_o}$ can be calculated. L_o, the inductance (at zero frequency) is evaluated from $L_o = \lim_{\omega \to o} \cfrac{X}{\omega}$. Using Equations (20) and (15) and letting $\omega \to o$,

$$L_o = \cfrac{4\mu \, \mu_r \, (\frac{b}{a})\sigma}{\pi^6 (\alpha_2-\alpha_1)^2 (\beta_2-\beta_1)^2 \sigma} \; \sum_{m=o}^{M} \sum_{k=o}^{M-m} \cfrac{\alpha^2(m)\beta^2(k)}{PP(m,k)} \tag{21}$$

and the ratio $\cfrac{L}{L_o}$ can be calculated for a range of frequencies. Equation (21) is of importance since it gives L_o, the inductance (at zero frequency), in closed form. Most techniques used for the solution of the type of problem considered here do not give an exact expression for L_o, the usual method being to evaluate it numerically for a sufficiently small value of ω.

Figures 2 and 3 show the variation of R/R_o , L/L_o with the frequency parameter αb for two insulation thicknesses and also for the case of the conductor completely filling the slot. The results obtained for this latter case agree with those obtained using the method described by Swann and Salmon (1963) for the fully open slot. From the graphs it can be seen that the effect of the insulation is to decrease the effective resistance and increase the effective inductance. When the insulation thickness is 5% of the slot dimensions, these effects are (when the frequency is given by $\alpha b = 6$)

(i) $\dfrac{R}{R_o}$ is decreased by 14% (approx)

(ii) $\dfrac{L}{L_o}$ is increased by 40% (approx).

If the insulation thickness is doubled, the graphs show that these changes are doubled.

5. CONCLUSIONS

The methods developed in this paper are capable of extension to a wide variety of similar practical problems. They can be used to give quantitative estimates of the effects of insulation for a range of frequencies and insulation thicknesses. The results of this study confirm that the presence of insulation does not greatly affect the effective impedance of the conductor.

7. REFERENCE

S.A. Swann, J.W. Salmon. Effective resistance and reactance of a

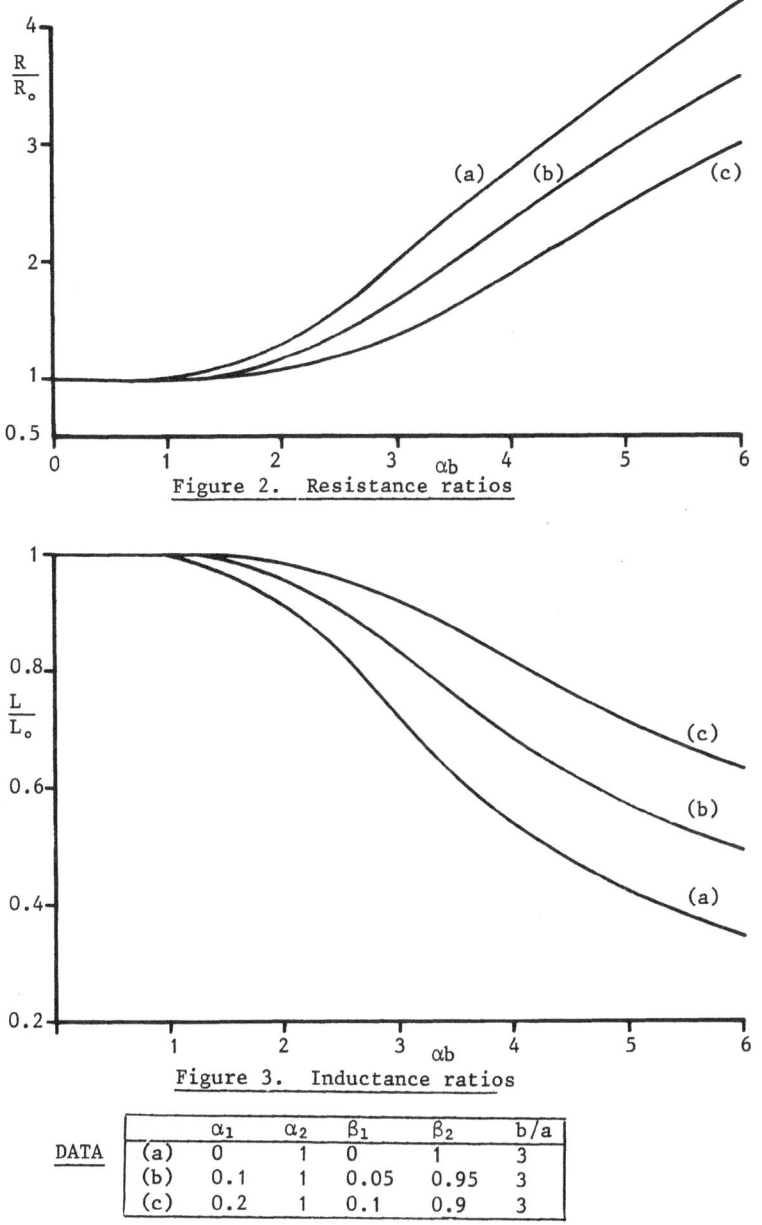

Figure 2. Resistance ratios

Figure 3. Inductance ratios

DATA		α_1	α_2	β_1	β_2	b/a
	(a)	0	1	0	1	3
	(b)	0.1	1	0.05	0.95	3
	(c)	0.2	1	0.1	0.9	3

rectangular conductor placed in a semi-closed slot. PROC.IEE, Volume
110, No. 9, pp 1656-1662, (1963).

APPENDIX A. NUMERICAL SUMMATION OF FOURIER SERIES

Now $\cos (n+1) \dfrac{\pi x}{\ell} - 2\cos \dfrac{\pi x}{\ell} \cos \dfrac{n\pi x}{\ell} + \cos (n-1) \dfrac{\pi x}{\ell} = 0$ (A1)

Define coefficients b_n, $n = 0, 1, \ldots (N+1)$ as follows:-

$\quad b_{N+1} = 0, \quad b_N = a_N$

$\quad b_n = a_n + 2\cos \dfrac{\pi x}{\ell} b_{n+1} - b_{n+2}$ for $n = (N-1), \ldots 1,0$ (A2)

then $\displaystyle\sum_{n=o}^{N}{}' \; a_n \cos \dfrac{n\pi x}{\ell} = \tfrac{1}{2} (b_0 - b_2)$ (A3)

Thus the series is summed by evaluating just one cosine term, namely
$2\cos\dfrac{\pi x}{\ell}$. The numerical calculation of the coefficients b_n, $n = 0,1,$
$\ldots$ (N+1) then follows trivially. Using the identity
$\cos(n+1+\tfrac{1}{2}) \dfrac{\pi x}{\ell} - 2\cos \dfrac{\pi x}{\ell} \cos(n+\tfrac{1}{2}) \dfrac{\pi x}{\ell} + \cos(n-1+\tfrac{1}{2}) \dfrac{\pi x}{\ell} = 0$ (A4)
and defining the coefficients b_n as above,

$\displaystyle\sum_{n=o}^{N} a_n \cos(n+\tfrac{1}{2}) \dfrac{\pi x}{\ell} = (b_0 - b_1) \cos \dfrac{\pi x}{2\ell}$

Notes. (1) Similar algorithms can be derived for Fourier sine series.
(2) To test the speed of the computing algorithms given here, the
following Fourier series was used.

$\displaystyle\sum_{n=o}^{N}{}' \; a_n \cos n\pi x + \sum_{n=1}^{N} A_n \sin n\pi x = \begin{cases} 1 & \text{for } \alpha \leqslant x \leqslant 1 \\ 0 & \text{for} -1 < x < \alpha \end{cases}$

giving $a_o = 1-\alpha$, $a_n = - \dfrac{1}{n\pi} \sin n\pi\alpha$ for $n = 1,2 \ldots N$,

$A_n = - \dfrac{1}{n\pi} \left\{ (-1)^n - \cos n\pi\alpha \right\}$ for $n = 1,2 \ldots N$

The series was evaluated for $x = -1(0.1)1$ using the algorithm and also
direct summation of the series. An internal computer procedure was
used to determine the actual computing time taken to sum the series in
each case. Using the algorithms the time taken was one quarter of that
used when the direct method of summation was employed. The savings
when evaluating a double Fourier series over a rectangular region are
even greater.

(3) When the algorithms given here are used in a computer, it is possible to overwrite the coefficients b_n as soon as each coefficient is no
longer needed. This is an advantage if computer storage is in short
supply.

APPENDIX B. SOLUTION OF A COUPLED SET OF EQUATIONS

$$PP(m,k) \; X_{mk} - \lambda \sum_{p=0}^{M} {}' \sum_{q=0}^{M-p} AA(p,m) \; BB(q,k) \; Y_{pq} = A(m) \; B(k) \tag{B1}$$

$$PP(m,k) \; Y_{mk} + \lambda \sum_{p=0}^{M} {}' \sum_{q=0}^{M-p} AA(p,m) \; BB(q,k) \; X_{pq} = 0 \tag{B2}$$

for $m = 0,1, \ldots M, \quad k = 0,1, \ldots (M-m)$

X_{mk}, Y_{mk} are the unknowns to be determined and $PP(m,k)$, $AA(m,k)$, $BB(m,k)$, $A(m)$, $B(k)$ and λ are known coefficients. Substituting for Y_{pq} from the Equation (B2) into (B1) gives

$$PP(m,k)X_{mk} + \lambda^2 \sum_{p=0}^{M} {}' \sum_{q=0}^{M-p} \frac{AA(p,m)BB(q,k)}{PP(p,q)} \sum_{i=0}^{M} {}' \sum_{j=0}^{M-i} AA(i,p)BB(j,q)X_{ij}$$

$$= A(m) \; B(k) \quad \text{for } m = 0,1, \ldots M, \quad k = 0,1 \ldots (M-m). \tag{B3}$$

Define arrays as follows:-

$$G_1(p,k) = \sum_{q=0}^{M-p} \frac{[BB(q,k)]^2}{PP(p,q)}, \; p = 0,1 \ldots M, \quad k = 0,1 \ldots M.$$

$$G_2(m,k) = \sum_{p=0}^{M} {}' [AA(p,m)]^2 \; G_1(p,k), \quad m = 0,1, \ldots M, \quad k = 0,1 \ldots M-m,$$

$$F_1(i,q) = \sum_{j=0}^{M-i} BB(j,q)X_{ij}, \quad i = 0,1 \ldots M, \quad q = 0,1 \ldots M$$

$$F_2(p,q) = \sum_{i=0}^{M} {}' AA(i,p) \; F_1(i,q), \; p = 0,1 \ldots M, \quad q = 0,1 \ldots (M-p)$$

$$F_3(p,k) = \sum_{q=0}^{M-p} \frac{BB(q,k)}{PP(p,q)} \; F_2(p,q), \; p = 0,1 \ldots M, \quad k = 0,1 \ldots M$$

$$F_4(m,k) = \sum_{p=0}^{M} {}' AA(p,m) \; F_3(p,k), \; m = 0,1 \ldots M, \quad k = 0,1 \ldots M-m$$

An iterative process for finding the unknowns X_{mk} is then given by

$$X_{mk}^{(n)} \left\{ PP(m,k) + \lambda^2 G_2(m,k) \right\} = A(m) \; B(k)$$
$$- \lambda^2 F_4^{(n-1)}(m,k) + \lambda^2 G_2(m,k)X_{mk}^{(n-1)} \; \text{for } m \neq 0 \tag{B4}$$

$$X_{ok}^{(n)} \left\{ PP(0,k) + \frac{\lambda^2}{2} G_2(0,k) \right\} = A(0)B(k) - \lambda^2 F_4^{(n-1)}(0,k)$$
$$+ \frac{\lambda^2}{2} G_2(0,k) \; X_{ok}^{(n-1)} \; \text{for } m = 0,1,2 \ldots M, \; k = 0,1,2 \ldots (M-m) \tag{B5}$$

$(X_{mk}^{(n)}$ denotes the value of X_{mk} after the n'th iteration.)

Having obtained the values of the coefficients X_{mk}, then the values of Y_{mk} are obtained from Equation (B2). It is not desirable to insert each new value of the unknown X_{mk} as it is found since this involves modification of the arrays F_1, F_2, F_3 and F_4 at each stage of the calculation. The method used calculates a complete set of iterates $X_{mk}^{(n)}$ for $m = 0,1 \ldots M$, $k = 0,1 \ldots (M-m)$ and then recalculates the arrays $F_1^{(n)}$, $F_2^{(n)}$, $F_3^{(n)}$ and $F_4^{(n)}$ ready for the next complete iteration.

SESSION B

APPLICATIONS & MODELLING TECHNIQUES II

Chairman

J CALDWELL

Newcastle upon Tyne Polytechnic

ELECTROMAGNETIC MODELING TECHNIQUES

USING FINITE ELEMENT METHODS

Z.J. Cendes

Electrical and Computer Engineering Department

Carnegie-Mellon University

Pittsburgh, PA 15213

1. INTRODUCTION

In designing and in evaluating finite element packages for industrial electromagnetics applications, the following questions are typical:

- How easy is the package to use?
- How accurate is the solution?
- Can I get smooth flux plots?
- What about the electric and magnetic fields?

The answers to these questions depend greatly on the techniques used to create the codes: a superior technique will result in fast, easy, and accurate models of industrial problems, while poor technique often means hours of set-up time to get jagged, inaccurate field approximations.

This paper describes some new techniques for finite element modeling by which the productivity, reliability and user acceptance of industrial electromagnetic CAD is greatly enhanced. The main ideas presented are (1) Delaunay triangulation, (2) direct vector field solution and (3) complementary solution error analysis.

2. DELAUNAY TRIANGULATION

Grid generation is often the most tedious aspect of finite element modeling. In the earliest grid generators, one entered the nodal coordinates and the node numbers specifying each triangle via punched cards. This procedure was very time consuming, required a great deal of effort, and was prone to human errors. A step forward from this is to use a linear or weighted interpolation method. In this method, the user is required to provide the end-point coordinates and the number of nodes along a line; the program generates the intermediate nodes from the end-point data by placing nodes at certain distances from each other. The majority of existing electromagnetics packages employ a variation of the interpolation method for finite element grid generation.[2]

There are two shortcomings with interpolatory grid generators:

41

1. It is difficult to make them work in all situations; and
2. The shape and distribution of the finite elements produced is often far from optimal.

To obtain accurate solutions, Babuska and Aziz[1] have proved that no angle in a triangular finite element grid may be close to 180°. (Note that often stated accuracy condition for finite element grids that no triangle should contain a small angle is wrong; what is essential for accurate solution is that no two of the three angles in a finite element grid be small). Here we propose to show that the triangulation defined in 1934 by the mathematician Delaunay[4] satisfies this finite element accuracy condition. In fact,the Delaunay triangulation algorithm provides an automatic way to insure that the triangles in a grid are maximally fat, and hence that a sound and accurate finite element grid is produced.

Given a set of points on a plane, the two main properties of the Delaunay triangulation of the N points are the following:

1. The Delaunay triangulation maximizes the sum of the minimum angles of all the triangles in the grid.
2. No circle drawn through the three vertices of any triangle in a Delaunay grid may include any other triangle vertex.

This second property of the Delaunay algorithm provides the following algorithm, due to Lawson[4], for its construction: Assume for the moment that the Delaunay triangulation of N points is known, and that we wish to add an (N+1)'st point inside one of the triangles. The added point provides a subdivision of the triangle into three smaller triangles; each of these new triangles forms a quadrilateral with its neighbor. Consider one of the three quadrilaterals and construct the circumcircle of one of the associated triangles. Since the circumcircle of a Delaunay triangle may not contain another Delaunay vertex in its interior, if the fourth point of the quadrilateral is located within the circle, the diagonal is swapped, otherwise no change in the data occurs. In the event that the quadrilateral diagonal is swapped, it is necessary to check the new triangles again for additional swaps. This procedure repeats until no more swaps are necessary.

Beginning with a simple grid of two triangles and four points which cover the entire graphics screen, it is possible to add points one at a time, applying the swapping procedure described above, until the complete finite element grid is generated. An example of a finite element grid obtained by using this procedure is given in Figure 1.

3. DIRECT ELECTRIC OR MAGNETIC FIELD SOLUTION

A large number of electromagnetic field problems satisfy the vector equations

$$\text{div } \bar{V} = \rho \qquad\qquad \text{curl } \bar{V} = \bar{J}$$

(1)

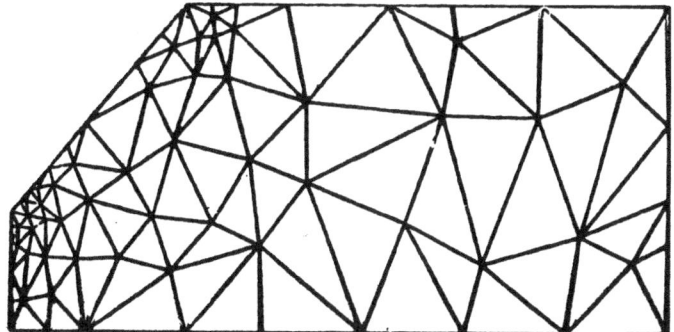

Figure 1: The Delauay triagulation of a notched conductor.

For example, in electrostatics the quantity $\bar{V}$ represents the electric field, ρ is the charge density divided by the permittivity, and $J = 0$. In magnetostatics, V is the magnetic field intensity, J is the current density times the permeability and $\rho = 0$.

Unfortunately, however, people do not usually solve equations Equation (1) directly but postulate solutions in terms of the scalar potential ϕ or in terms of the vector potential A

$$\bar{V} = - \text{grad } \phi \qquad \text{or} \qquad \bar{V} = \text{curl } \bar{A}$$

$$(2)$$

The scalar potential ϕ is used if $\bar{J} = 0$, while the vector potential $\bar{A}$ is used if $\rho = 0$. The disadvantages of these potential formulations is the fact that the field $\underline{V}$ is expressed in terms of the derivatives of the potentials ϕ and A   Since ϕ and A are computed only approximately in numerical solutions, and since numerical differentiation is highly inaccurate, the field values evaluated by using this procedure are also highly inaccurate.

Here we show a technique for modeling electromagnetic field problems directly in terms of the electric and magnetic fields.   For the same matrix size, the field values thus computed are an order of magnitude more accurate than computations derived from potential methods.

3.1. DIRECT EXPRESSIONS

It is relatively easy to compute the divergence and curl of a given finite element vector V, from universal matrices published for the directional derivative.[5]

To begin, we define homogeneous coordinates $\{ \zeta_\ell, \quad \ell = 1,2,3, \}$ on an arbitrary triangle by the usual affine transformation

$$\begin{bmatrix} 1 \\ x \\ y \end{bmatrix} = \begin{bmatrix} 1 & 1 & 1 \\ x_1 & x_2 & x_3 \\ y_1 & y_2 & y_3 \end{bmatrix} \begin{bmatrix} \zeta_1 \\ \zeta_2 \\ \zeta_3 \end{bmatrix}$$

(3)

where $\{(x_\ell, y_\ell), \ell = 1, 2, 3\}$ are the three sets of triangle vertex coordinates. As is well known[11], this transformation is invertible, giving the homogeneous coordinates $\{\zeta_\ell\}$ in terms of the point coordinates (x,y)

$$\begin{bmatrix} \zeta_1 \\ \zeta_2 \\ \zeta_3 \end{bmatrix} = \frac{1}{\Delta} \begin{bmatrix} a_1 & b_1 & c_1 \\ a_2 & b_2 & c_2 \\ a_3 & b_3 & c_3 \end{bmatrix} \begin{bmatrix} 1 \\ x \\ y \end{bmatrix}$$

(4)

where with (ℓ, m, n) cyclic modulo three

$$a_\ell = x_m y_n - x_n y_m$$
$$b_\ell = y_m - y_n \qquad \qquad \Delta = \sum_{\ell=1}^{3} (x_\ell y_m - x_\ell y_n)$$
$$c_\ell = x_n - x_m$$

(5)

The x and y derivatives of a function $\phi(x,y)$ may be expressed in terms of the homogeneous coordinates ζ_ℓ by using the chain rule and equation (4)

$$\frac{\partial \phi}{\partial x} = \sum_{\ell=1}^{3} \frac{\partial \zeta_\ell}{\partial x} \frac{\partial \phi}{\partial \zeta_\ell} = \frac{1}{\Delta} \sum_{\ell=1}^{3} b_\ell \frac{\partial \phi}{\partial \zeta_\ell}$$

(6)

with a similar expression for $\partial \phi / \partial y$. Let $a_i^{(n)}(x,y)$ be the n'th order interpolation polynomial defined by Silvester[5] and approximate $\phi(x,y)$ on a triangular finite element by the sum

$$\phi(x,y) = \tilde{a}^{(n)}(x,y) \, \phi$$

(7)

where $\tilde{a}^{(n)}(x,y)$ is a row vector with elements $a_i^{(n)}$ and ϕ is a column vector with elements ϕ_i.

44

The derivative of $\tilde{a}^{(n)}(x,y)$ is $(n-1)$'st order and is given by

$$\frac{\partial a^{(n)}}{\partial \zeta_\ell} = \tilde{a}^{(n-1)}G_\ell \tag{8}$$

where G_ℓ is a numerical matrix independent of triangle shape and size and is published in reference 5 for $n = 1,...,4$. Substituting equations (7) and (8) into equation (6) gives

$$\frac{\partial \phi}{\partial x} = a^{(n-1)}D_x\ell \qquad \text{where} \qquad D_x = \frac{1}{\Delta} \sum_{\ell=1}^{3} b_\ell \, G_\ell \tag{9}$$

A similar equation, with b_ℓ replaced by c_ℓ, is obtained for $\partial\phi/\partial y$. Note that the dimension of the matrix D_x is $r = n(n+1)/2$ by $c = (n+1)(n+2)/2$.

3.2. MATRIX REPRESENTATIONS

For two component vectors, the divergence equation (1a) becomes

$$\frac{\partial V_x}{\partial x} + \frac{\partial V_y}{\partial y} = \rho \tag{10}$$

Approximating V_x and V_y by n'th order polynomials, using equation (15) and evaluating both sides of this equation at the interpolation nodes yields

$$\begin{bmatrix} D_x & D_y \end{bmatrix} \begin{bmatrix} \underset{\sim}{V}_x \\ \underset{\sim}{V}_y \end{bmatrix} = [\underset{\sim}{\rho}] \tag{11}$$

Thus the divergence operator acting on a triangular finite element is equivalent to the matrix

$$\text{Div} = \begin{bmatrix} D_x & D_y \end{bmatrix} \tag{12}$$

Note that the dimension of the nullspace of this matrix is $\eta(\text{Div}) = 2N - M = (n+1)(n+4)/2$.

45

In two dimensions, the curl equation is similar to the divergence equation and produces nearly the same result. Provided that the vector V is invariant in the z-direction and has only x and y components the curl operator acting on a triangular finite element is seen to be equivalent to the matrix

$$\text{Curl} = \begin{bmatrix} -D_y & D_x \end{bmatrix}$$

(13)

3.3. SOLUTIONS
In view of the above results, we conclude that the combined finite element matrix representation of equation (1) is

$$\begin{bmatrix} D_x & D_y \\ -D_y & D_x \end{bmatrix} \begin{bmatrix} V_x \\ V_y \end{bmatrix} = \begin{bmatrix} \rho \\ J \end{bmatrix}$$

(14)

Note that equation (14) expresses the divergence and curl conditions on $\overline{V}$ exactly under the assumption that V is approximated by a polynomial. A general result, due to Lee and Schachter, is that the number of triangles T and the number of edges E in a triangle mesh is given by

$$T = 2(V-1) - B \qquad E = 3(V-1) - B$$

(15)

where V is the total number of vertices in the mesh and B is the number of vertices on its exterior boundary. Assuming that the mesh consists of n'th order finite elements, there are the V vertex points, plus $(n-1)$ points on each edge and $(n-2)(n-1)/2$ points in the interior of each element. Thus, the total number of unknowns U in a two-component vector solution is

$$U = 2V + 2(n-1)E + (n-2)(n-1)T$$

(16)

Using equation (15) gives

$$U = n^2 T + nB + 2$$

(17)

Of course, the number of equations W derived from equation (14) for the T finite elements is

46

$$W = n(n+1)T \tag{18}$$

Subtracting equation (18) from (17) gives the difference

$$R = W - U = n(T - B) - 2 \tag{19}$$

Since R is a positive number in all practical cases, we see that, despite the fact that equation (14) for one element is underdetermined, global finite element matrices based on equation (14) are overdetermined. Adding boundary conditions to this global matrix simply makes the system even more overdetermined.

One method of solving overdetermined systems of equations is to compute a least-squares solution by multiplying both sides by the transpose of the coefficient matrix. This is the approach taken in reference 3. A typical direct magnetic field vector solution obtained by using this procedure is shown in Figure 2.

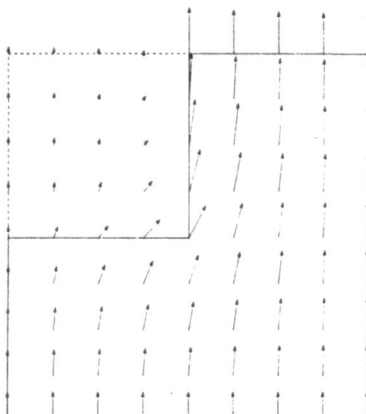

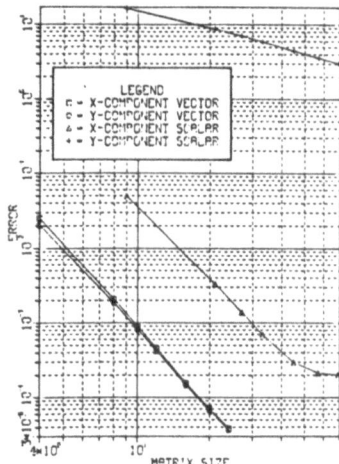

Figure 2: The Magnetic field vector obtained by least-squares solution for an electric machine slot.

Figure 3: Comparison of the accuracies of direct electric field solution and scalar potential solution for a circular charge cylinder

To compare the accuracy of solutions obtained by using the direct vector approach with that obtained by using the traditional scalar finite element method, several problems having analytical solutions were solved with both procedures. Figure 3 presents the results from one of the studies. In this case, the electric field

47

produced by a charged circular cylinder is computed, the exact solution of which is known to decay as $1/r$ outside the cylinder. A comparison of the accuracy of the electric field values obtained from the scalar and vector solution procedures reveals that the vector procedure is more than an order of magnitude more accurate than the traditional scalar approach.

4. ERROR ANALYSIS

Since the global finite element equation derived from equation (14) is overdetermined, we have a choice. We can compute its least-squares solution as described above, or we can uncouple the system of equations and solve either the divergence or the curl equation exactly, throwing all of the error into approximating the remaining equation.

4.1. THE DIVERGENCE

To solve the divergence equation exactly, let us first rewrite equation (10) in the homogeneous form

$$\rho = Z\, F^{-1}V \quad \text{where} \quad Z = \left[\frac{\partial}{\partial \zeta_1} \quad \frac{\partial}{\partial \zeta_2} \quad \frac{\partial}{\partial \zeta_3}\right]$$

(20)

and

$$F^{-1} = \frac{1}{\Delta} \begin{bmatrix} a_1 & b_1 & c_1 \\ a_2 & b_2 & c_2 \\ a_3 & b_3 & c_3 \end{bmatrix} \qquad V = \begin{bmatrix} 0 \\ v_x \\ v_y \end{bmatrix}$$

(21)

To make the analysis simpler, let

$$U = F^{-1}V$$

(22)

and approximate each component of U by n'th order interpolation polynomials

48

$$U = \left[I_3 \otimes \tilde{a}^{(n)} \right] \underset{\sim}{U} \tag{23}$$

where I_3 is the 3 by 3 identity matrix and $\otimes$ denotes the Kronecker product. Equation (20) then becomes

$$\rho = \tilde{a}^{(n-1)} D \underset{\sim}{U} \quad \text{where} \quad D = [G_1 \quad G_2 \quad G_3] \tag{24}$$

Approximating ρ in terms of $(n-1)$'st order polynomials as in equation (11) yields

$$D \underset{\sim}{U} = \underset{\sim}{\phi} \tag{25}$$

By construction, the matrix G_1 is of full row rank so that the matrix D may be decomposed as

$$D = [L][U_1 \quad U_2] \tag{26}$$

where L is r by r and lower triangular, U_1 is r by r and upper triangular and U_2 is r by c and has no particularly useful structure. It follows that the general solution of equation (25) is

$$\underset{\sim}{U} = D \underset{\sim}{\phi} + N \underset{\sim}{w} \tag{27}$$

where the $(3c-r)$ components of $\underset{\sim}{w}$ are arbitrary and

$$D^+ = \begin{bmatrix} U_1^{-1} \ L^{-1} \\ 0 \end{bmatrix} \qquad N = \begin{bmatrix} -U_1^{-1} \ U_2 \\ L \qquad I \end{bmatrix} \tag{28}$$

Substituting equations (23) and (27) into equation (22) gives

$$V = F \ [I_3 \otimes \tilde{a}]\{D^+ \ \rho + N \underset{\sim}{w}\} \tag{29}$$

49

By definition, the first component of V is zero. Therefore, we must have that

$$\{[1 \ 1 \ 1] \otimes \tilde{a}^{(n-1)}\}\{D^+ \underset{\sim}{\rho} + N\underset{\sim}{w}\} = 0 \tag{30}$$

This may be rewritten as

$$\tilde{a}^{(n-1)}[(D_1^+ \underset{\sim}{\rho} + N_1\underset{\sim}{w}) \ (D_2^+ \underset{\sim}{\rho} + N_2\underset{\sim}{w}) \ (D_3^+ \underset{\sim}{\rho} + N_3\underset{\sim}{w})] \begin{bmatrix} 1 \\ 1 \\ 1 \end{bmatrix} = 0 \tag{31}$$

where D_1 and N_1 represent the first n rows of D and N, respectively, D_2 and N_2 the second n rows, and D_3 and N_3 the third. Noting that D_2^+ and D_3^+ are always zero, equation (31) becomes

$$\tilde{a}^{(n-1)}\left\{ D^+ \underset{\sim}{\rho} + \sum_{\ell=1}^{3} N_\ell \underset{\sim}{w} \right\} = 0 \tag{32}$$

It follows that

$$S \underset{\sim}{w} = -D^+ \underset{\sim}{\rho} \qquad \text{where} \qquad S = \sum_{\ell=1}^{3} N_\ell \tag{33}$$

Solving this gives

$$\underset{\sim}{w} = -S^- D \underset{\sim}{\rho} + N_s \underset{\sim}{z} \tag{34}$$

where $\underset{\sim}{z}$ is an arbitrary vector of length (2c-r). Substituting (34) into (29) yields

$$V = F[I_3 \otimes \tilde{a}]\{R \underset{\sim}{\rho} + M \underset{\sim}{z}\}$$
where $\tag{35}$
$$R = (I_{3c} - N S^+) D^+$$

$$M = NN_s \tag{36}$$

50

The second and third components of V are V_x and V_y, which become

$$\begin{bmatrix} V_x \\ V_y \end{bmatrix} = \begin{bmatrix} x_1 & x_2 & x_3 \\ y_1 & y_2 & y_3 \end{bmatrix} \otimes \tilde{a} \left\{ R\tilde{\rho} + Mz \right\} \tag{37}$$

Finally, evaluating equation (37) on both sides at the interpolation nodes gives the desired result

$$\underset{\sim}{\bar{V}} = B_n R \underset{\sim}{\rho} + B_n M \underset{\sim}{z} \tag{38}$$

where

$$B_u = \begin{bmatrix} x_1 & x_2 & x_3 \\ y_1 & y_2 & y_3 \end{bmatrix} \otimes I_n \tag{39}$$

Note that the matrices R and M are independent of triangle coordinates and that the matrix M provides $(2c-r) = (n+1)(n+4)/2$ nullvectors for V, as required.

By plugging in numerical values, it follows that equation (38) becomes for first order elements

$$\underset{\sim}{V}_x = \begin{bmatrix} -c_3 \\ 0 \\ 0 \end{bmatrix} [\rho] + \begin{bmatrix} c_3 & 0 & c_1 & 0 & c_3 \\ c_3 & 0 & 0 & -c_2 & 0 \\ 0 & c_3 & 0 & 0 & -c_2 \end{bmatrix} \underset{\sim}{z}$$

$$\underset{\sim}{V}_y = \begin{bmatrix} b_3 \\ 0 \\ 0 \end{bmatrix} [\rho] + \begin{bmatrix} -b_3 & 0 & -b_1 & 0 & -b_3 \\ -b_3 & 0 & 0 & b_2 & 0 \\ 0 & -b_3 & 0 & 0 & b_2 \end{bmatrix} \underset{\sim}{z} \tag{40}$$

The reader may check that taking the divergence of a vector with the above components indeed has divergence ρ. A similar result is found for curl $\tilde{V}$.

4.2. ADAPTIVE SOLUTION PROCEDURE

Let $V^{(1)}$ be the solution obtained by solving equation (1a) exactly for $\bar{V}$ while approximating the conditions in equation (1b). Similarly, let $V^{(2)}$ be the solution obtained by solving equation (1b) exactly but approximating equation (1a). It can be shown[2] that the energy norm of the difference between $V^{(1)}$ and $V^{(2)}$ is an upper bound on the error in either $V^{(1)}$ or $V^{(2)}$.

The above principle leads to the following adaptive procedure[2] for the solution of electromagnetic field problems: Given any field problem geometry, subdivide this geometry into the minimum number of finite elements possible; use the Delaunay algorithm to generate an optimal triangulation. Solve the problem with this minimal grid twice, once solving equation (1a) exactly, once solving equation (1b) exactly, as described in the preceding sections. By computing the difference

51

between the two complementary solutions, an upper bound to the error in the solution sense is produced. Subdivide those elements with an error greater than a specified amount by introducing a new node at the the element centroids; once again use the Delaunay algorithm to generate the optimal triangulation. Resolve and recompute the errors, and refine the mesh if needed, until the error bound in all elements is within the specified tolerance.

The vertex points in the Delaunay in Figure 1 were evaluated by using the above procedure; Figure 4 shows a portion of the solution obtained. Note that as expected the finite element mesh produced by the adaptive procedure is finest where the field varies most rapidly.

Figure 4   Equipotential contours and streamlines for a portion of the problem in Figure 1.

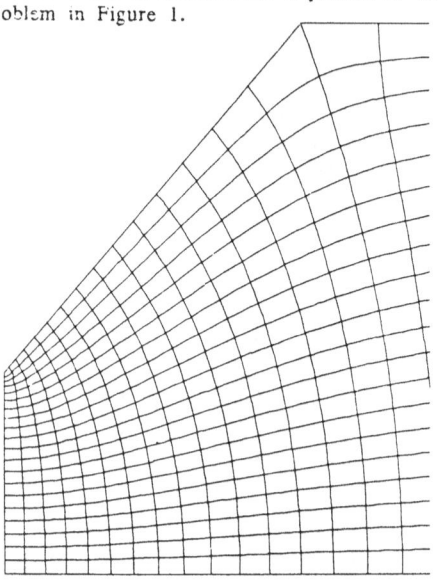

REFERENCES

1. Babuska, I., and Aziz, A.K., "On the Angle Condition in the Finite Element Method", SIAM J. Numerical Anal., pp. 214-216, 1976.
2. Cendes, Z.J., Shenton. D., and Shahnasser. H., "Magnetic Field Computation using Delaunay Triangulation and Complementary Finite Element Methods", submitted to IEEE Transactions on Magnetics.
3. Hoole, S.R.H., Direct Finite Element Solution of the Magnetic Field Vector, Ph.D. Thesis, Carnegie-Mellon University, Pittsburgh, PA, 1983.
4. Lee, D.T., and Schacter, B.J., "Two Algorithms for Constructing a Delaunay Triangulation", Int. J. of Comp. and Inf. Sci., Vol. 9, No. 3, pp. 219-241, 1980.
5. Silvester, P., and Haslam, C.R.S., Magnetotelluric Modeling by the Finite Element Method", Geophysical Prospecting, Vol. 20, pp. 872-891, 1972.

ERROR BOUNDED FORMULATIONS IN ELECTROMAGNETISM

J. Penman[1] and J.R. Fraser[2]

1. Department of Engineering, University of Aberdeen
2. McDermott Engineering Ltd.

ABSTRACT

This paper seeks to show, in a relatively general way, that it is possible to choose different formulations of electromagnetic field problems in such a way that the variational solutions obtained are error bounded. This makes the technique suitable for use with the finite element method, and can lead to substantial economies in computer usage.

1. INTRODUCTION

The solution of problems in applied electromagnetics, using numerical techniques such as the finite element method, is now carried out routinely and often interactively. This is due principally to the recent increases in computational power and advances in pre- and post-processing techniques. Large scale problems still stretch the resources of many computer installations, however. It is therefore of great advantage to know when a given problem is being solved in the most efficient and convenient way. Just such problems arise with the finite element method for unless the user chooses to examine successive solutions to a problem, as a function of the number of degrees of freedom used, then it is difficult to get any impression of convergence or absolute accuracy. Adopting such a course, however, is time consuming and expensive, both in terms of computer usage, and data preparation.

Arthurs (1972) has examined the possibility of complementary variational formulations for a wide class of problems. More recently Hammond and Penman (1976, 1978) developed such methods and applied them to simple static and eddy current systems. The authors of this paper have extended this work and implemented it in finite element form (Penman and Fraser, 1982, 1983). These recently developed dual and complementary techniques allow a choice of problem formulation and lead to an indication of error. The purpose of this paper is to illustrate the salient features of the method when applied to problems in electromagnetism.

53

2. THE GENERAL FIELD PROBLEM

We will consider problems that can be defined as

    (i) a compatibility equation
   (ii) a constitutive equation
  (iii) an equilibrium equation.

These terms are borrowed from the field of stress analysis but their meaning carries over to other areas, as we shall see.

Following Tonti (1972) such a framework can be described diagrammatically, as in Figure 1.

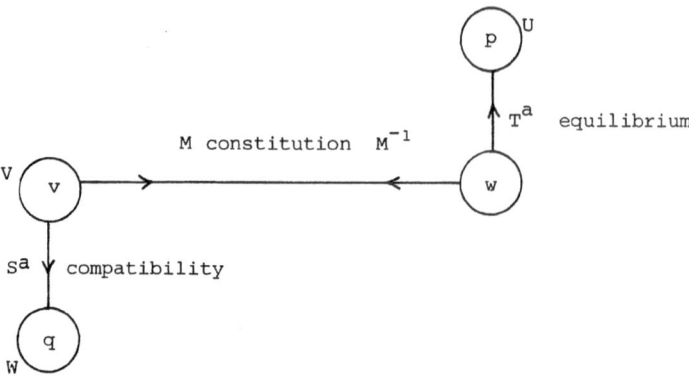

Figure 1

Two further functions u ε U, and v_s ε V are introduced, together with the operator T so that,

if $v = v_s + Tu$ (1)

then $S^a v = S^a v_s + S^a Tu$

             $= q + 0$

i.e. if equation (1) is satisfied, the compatibility relationship is automatically satisfied. v_s is the source component of v and is such that when operated on by S^a it gives the function q exactly. The range of T is the null space of S^a, so in the present context $S^a Tu = 0$ for all u in the domain of T.

Similarly a further two functions are introduced, r ε W and w_s ε V,

54

together with the operator S so that,

if $w = w_s + Sr$ (2)

then $T^a w = T^a w_s + T^a Sr$

       $= p + 0$

i.e. if equation (2) is satisfied then the equilibrium relationship
is automatically satisfied. w_s is the source component of w, and when
operated upon by T^a yields the function p exactly. Similarly to T,
the range of the operator S is the null space of T^a. All of these
relationships are represented diagrammatically in Figure 2.

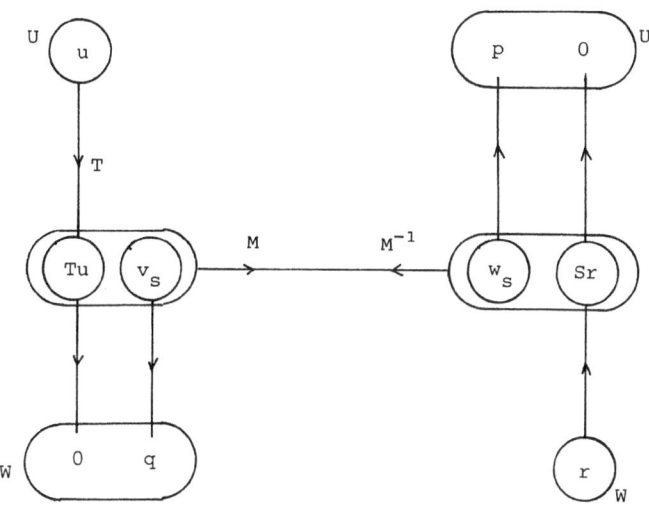

Figure 2

We consider three kinds of boundary condition,

    (i) the far boundary
   (ii) compatibility boundary condition, $E^* v = \bar{h}$
  (iii) equilibrium boundary condition, $B^* w = h$

The far boundary condition implies a boundary remote from the sources
of field, hence field values of zero are appropriate. Now if the
compatibility equation is indirectly satisfied as in equation (1)
then the corresponding boundary condition, given above can be

indirectly stated by requiring that,

$$Bu = g \quad \text{on } \partial\Omega_1 \tag{3}$$

Similarly if equilibrium is indirectly satisfied as in equation (2) then the equilibrium boundary condition is indirectly stated as,

$$Er = \bar{g} \quad \text{on } \partial\Omega_2 \tag{4}$$

The far boundary condition, if appropriate, can be considered a special case of (3) and (4) with g, or $\bar{g}$, equal to zero.

We may thus proceed in two ways. The first is to indirectly satisfy compatibility and the corresponding boundary condition, then satisfy the constitutive relationship, and the equilibrium equation and boundary condition. This requires that,

$$Tu + v_s = v \qquad \text{with } Bu = g \text{ on } \partial\Omega_1$$

$$Mv = w \qquad \text{in } \Omega \tag{5}$$

$$T^a w = p \qquad \text{with } B^* w = h \text{ on } \partial\Omega_2$$

The alternative is to indirectly satisfy equilibrium and the associated boundary condition, then directly satisfy compatibility and constitution. This leads to

$$S^a v = q \qquad \text{with } E^* v = \bar{h} \quad \text{on } \partial\Omega_1$$

$$v = M^{-1} w \qquad \text{in } \Omega \tag{6}$$

$$Sr + w_s = w \qquad \text{with } Er = \bar{g} \quad \text{on } \partial\Omega_2$$

The behaviour of the system can therefore be described by two different canonical sets of equations, (5) and (6).

If we define w to be $M(Tu + v_s)$ then (5) becomes,

$$T^a M(Tu + v_s) = p \quad \text{in } \Omega \tag{7}$$

with $Bu = g$ on $\partial\Omega_1$ and $B^* w = h$ on $\partial\Omega_2$.

Similarly if v is defined as $M^{-1}(Sr + w_s)$ then (6) gives

$$S^a M^{-1}(Sr + w_s) = q \text{ in } \Omega \tag{8}$$

with $E^* v = \bar{h}$ on $\partial\Omega_1$ and $Er = \bar{g}$ on $\partial\Omega_2$.

We call equation (7) the PRIMAL form for the problem, whilst equation

(8) is termed the DUAL form. It has been shown elsewhere (Penman and Fraser, 1982) that it is possible to obtain two complementary functionals to solve equation (7). Since equation (8) has exactly the same form a further two complementary functionals can be obtained to solve it. This means there are four possible functionals from which a solution to the original field problem can be obtained. They are,

(i) $\theta(u)$, which minimises to a weak solution to
$T^a w = p$ in Ω with $B^* w = h$ on $\partial\Omega_2$, provided
that $w = Mv$, $Tu + v_s = v$ in Ω, and
$Bu = g$ on $\partial\Omega_1$.

This is the STANDARD PRIMAL METHOD.

(ii) $\Xi(w)$, which maximises to a weak solution to
$Tu + v_s = v$ in Ω with $Bu = g$ on $\partial\Omega_1$, provided
that $w = Mv$, $T^a w = p$ in Ω, and $B^* w = h$ on $\partial\Omega_2$.

This is the COMPLEMENTARY PRIMAL METHOD.

(iii) $\alpha(r)$, which minimises to a weak solution to
$S^a v = q$ in Ω with $E^* v = \bar{h}$ on $\partial\Omega_1$, provided
that $w = Mv$, $Sr + w_s = w$ in Ω, and $Er = \bar{g}$
on $\partial\Omega_2$.

This is the STANDARD DUAL METHOD.

(iv) $\beta(v)$, which maximises to a weak solution to
$Sr + w_s = w$ in Ω with $Er = g$ on $\partial\Omega_2$ provided
that $w = Mv$, $Sr + w_s = w$ in Ω, and $E^* v = \bar{h}$
on $\partial\Omega_1$.

This is the COMPLEMENTARY DUAL METHOD.

Fraser (1982) has also shown that when approximating the functions u and w by u_a and w_a that,

$$\theta(u_a) \geq \theta(u) = \Pi(u,w) = \Xi(w) \geq \Xi(w_a) \tag{9}$$

Similar boundedness exists for $\alpha(r_a)$ and $\beta(v_a)$, where r_a and v_a are approximations to the functions r and v.

3.   THE MAGNETOSTATIC FIELD

The general problem outlined above can be easily simplified to represent the forms usually associated with the electromagnetic field. Here we illustrate this by considering the magnetostatic field.

The magnetostatic field requires that the following equations be satisfied in the domain Ω.

$$\nabla \cdot \underline{B} = 0 \quad - \quad \text{compatibility}$$

$$\underline{B} = \mu\underline{H} \quad - \quad \text{constitutive} \tag{10}$$

$$\nabla \times \underline{H} = \underline{J} \quad - \quad \text{equilibrium}$$

The compatibility boundary condition is,

$$\underline{n} \cdot \underline{B} = \bar{h} \quad \text{on } \partial\Omega_1 \quad - \quad \text{compatibility, and}$$

$$\underline{n} \times \underline{H} = \underline{h} \quad \text{on } \partial\Omega_2 \quad - \quad \text{equilibrium.}$$

If we define $\underline{B} = \nabla \times \underline{A}$, where $\underline{A}$ is the magnetic vector potential then $\nabla \cdot \underline{B} = \nabla.\nabla \times \underline{A} = 0$, so that the compatibility equation, in Ω, is satisfied. The corresponding boundary condition becomes, $\underline{n}.\nabla \times \underline{A} = \bar{h}$ on $\partial\Omega_1$ which is satisfied if $\underline{A} = \underline{g}$ on $\partial\Omega_1$.

The set of equations given by (10) thus lead to the canonical set,

$$\nabla \times \underline{A} = \underline{B} \qquad \text{with } \underline{A} = \underline{g} \quad \text{on } \partial\Omega_1$$

$$\underline{H} = \frac{1}{\mu} \underline{B} \qquad \text{in } \Omega \tag{11}$$

$$\nabla \times \underline{H} = \underline{J} \qquad \text{with } \underline{n} \times \underline{H} = \underline{h} \quad \text{on } \partial\Omega_2$$

Comparing equations (5) and (11), and relating the magnetostatic field vectors to those denoted in Figures 1 and 2 it can be seen that the following equivalences hold:

$$\underline{A} \equiv u \, , \qquad T \equiv \nabla \times$$
$$0 \equiv v_s , \qquad T^a \equiv \nabla \times$$
$$\underline{J} \equiv p \, , \qquad S \equiv -\nabla$$
$$\underline{H} \equiv w \, , \qquad S^a \equiv \nabla .$$

Note also that we let $\underline{H} = \underline{H}_s - \nabla\phi$, where $\underline{H}_s \equiv w_s$ and $\phi \equiv r$. That is $\underline{H}$ is composed of a scalar derived part and a vector derived part. This, of course, is commonly done.

The canonical set, equation (11), defines the PRIMAL form,

$$\nabla \times \frac{1}{\mu} \nabla \times \underline{A} = \underline{J}, \quad \text{as expected.}$$

To obtain bounded solutions it is therefore necessary to extremise the functionals, $\theta(\underline{A})$ and $\Xi(\underline{H})$. These are the standard primal, and

58

standard complementary forms. The dual forms can easily be identified in the same way. The authors have shown (Penman and Fraser, 1982) that,

$$\theta(\underline{A}) = \tfrac{1}{2} \int_{\Omega} (\nabla \times \underline{A}) \cdot \frac{1}{\mu} (\nabla \times \underline{A}) d\Omega - \int_{\Omega} \underline{J}.\underline{A} \, d\Omega + \int_{\partial\Omega_2} \underline{h}.\underline{A} \, d(\partial\Omega) \tag{12}$$

with $\underline{A} = \underline{g}$ specified on $\partial\Omega_1$, and

$$\Xi(\underline{H}) = -\tfrac{1}{2} \int_{\Omega} \underline{H}.\mu\underline{H} \, d\Omega - \int_{\partial\Omega_1} \underline{g}.(\underline{n} \times \underline{H}) \, d(\partial\Omega) \tag{13}$$

with $\underline{n} \times \underline{H} = \underline{h}$ on $\partial\Omega_2$.

Also, as in equation (9),

$$\theta(\underline{A}_a) \geq \Pi(\underline{A},\underline{H}) \geq \Xi(\underline{H}_a) \tag{14}$$

It is computationally more efficient to write equation (13) in an alternative form, generated by substituting $\underline{H}_s - \nabla\phi$ for $\underline{H}$.

This yields,

$$\Xi(\phi) = -\tfrac{1}{2} \int_{\partial\Omega} (\underline{H}_s - \nabla\phi).\mu(\underline{H}_s - \nabla\phi) d\Omega - \int_{\partial\Omega_1} \underline{g}.\underline{n} \times (\underline{H}_s - \nabla\phi) \, d(\partial\Omega) \tag{15}$$

with $\phi = \bar{g}$, specified on $\partial\Omega_2$.

The two functionals of equations (12) and (15) can be extremised using the finite element method in the usual way, and provide bounds on the energy of the system, which can be related to the quantity $\tfrac{1}{2}Li^2$. Here L is inductance and i, current.

This ability to generate global error bounds allows significant computational advantage to be gained. This is principally because the average of the two bounded solutions, obtained using a modest number of degrees of freedom, can achieve an accuracy normally only obtainable using a large number of degrees of freedom with the normal finite element formulation. This is adequately illustrated by the simple problem shown below.

4.   RESULTS

The problem to be solved is specified in Figure 3 and the upper and lower bounds on the system energy, as a function of number of degrees of freedom, given in Figure 4.

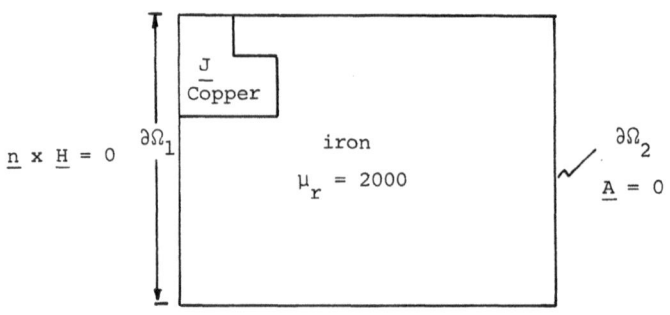

Figure 3

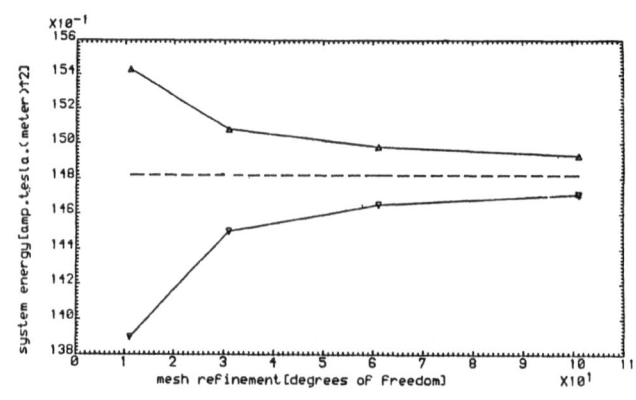

Figure 4

5. CONCLUSIONS

Dual and complementary variational formulations for a wide class of field problems have been developed, and it has been shown that error bounded solutions can be generated using the finite element method. The technique is extremely general and can lead to large savings in

60

computer usage. When dealing with 2-dimensional problems savings of approximately 20:1 are easily achievable when calculating global quantities. It is also observed that the bounded value of the local field variable is more accurate than either of the bounding values.

6.   NOMENCLATURE

V, U, W   vector spaces

v, u, w
p, q, r   vectors from the above spaces
v_s, w_s

T, T^a
S, S^a linear operators and the corresponding adjoints
M

B, B^*, E, E^* boundary operators
Ω problem domain
$\partial\Omega$ boundary surface
$\theta, \Xi, \alpha, \beta, \Pi$ functionals
$\underline{B}$ magnetic flux density
$\underline{H}$ magnetic intensity
$\underline{J}$ current density
$\underline{A}$ magnetic vector potential
ϕ magnetic scalar potential
μ permeability

7.   REFERENCES

1.   Arthurs, A.M. 'Complementary variational principles', published O.U.P., 1970.

2.   Fraser, J.R. 'Complementary and dual finite element principles', Ph.D. Thesis, Department of Engineering, University of Aberdeen, 1982.

3.   Hammond, P., Penman, J. 'Calculation of inductance and capacitance by means of dual energy principles', Proc. IEE, Vol. 123, No. 6, 1976, pp 554-559.

4.   Hammond, P., Penman, J. 'Calculation of eddy currents by dual energy methods', Proc. IEE, Vol. 125, No. 7, 1978, pp 701-708.

5.   Penman, J., Fraser, J.R., 'Complementary and dual energy finite element principles in magnetostatics', Trans. IEEE (Mag), Vol. 18, No. 2, 1982, pp 319-324.

61

6. Penman, J., Fraser, J.R. 'The efficient calculation of electro-
    static fields in large systems', Electrostatics 83, pub. Inst.
    Phys., 1983, pp 243-248.

7. Tonti, E. 'On a mathematical structure of a large class of
    physical theories', Acad. Naz. dei Lineei, Vol. LII, Sine III,
    1972, pp 48-56.

SESSION C

APPLICATIONS & MODELLING TECHNIQUES III

Chairman

T S WILKINSON

NEI Parsons

ELECTROMAGNETIC MODELLING TECHNIQUES USING BOUNDARY ELEMENT METHODS

A. Wexler

University of Manitoba, Winnipeg, Canada

ABSTRACT

This paper describes how Langrangian and spline methods are used in high-order boundary element method (BEM) codes. These methodologies permit high-fidelity modelling of both source distributions and surfaces and yield improved efficiency and accuracy. A brief description of a new BEM program is included.

1. INTRODUCTION

The boundary element method (BEM) consists of a set of modular algorithms and embodies the integral equation method in the solution of field problems. In its modularity and efficiency, the BEM resembles the finite element method (FEM) which employs the partial differential equation formulation. In the BEM, unknown variables – rather than being potentials at nodes dispersed throughout a volume – can be represented as sources located solely on boundaries and interfaces. Essentially, the dimensionality of a problem is reduced: e.g. a three-dimensional distribution of nodes is replaced by a surface distribution.

When properly implemented, the BEM can be organized in such a fashion that it borrows heavily from the FEM. For example, the isoparametric method using Lagrange shape functions can also be used in the BEM in order to model curved surfaces as curved surfaces rather than in a sequence of steps or in a piecewise planar fashion. Thus, a surface can be modelled as a number of non-planar triangles, say, connected at edges. This is effected by using quadratic (or higher-order) shape functions. However, although surface continuity is ensured in this way, Lagrangian element surface modelling is subject to derivative discontinuities across common element boundaries. Therefore incorrect and large local fields can exist in the vicinity of these accidental creases.

A preferred modelling technique is the spline method - in particular, its implementation in the Coons patch technique. A cubic spline exhibits second derivative continuity and, by using several spline functions, a Coons patch surface may be defined. Such patches may be used as the basis of a BEM method. It turns out that considerable storage and computational economies are realized because of the

65

sharing of information between adjacent elements. Coons patch modelling can also serve as the basis of a FEM mesh-generating scheme.

2. THE INTEGRAL EQUATION FORMULATION

The integral equation (IE) method employs source distributions in order to determine the field at any distance. For simplicity, the use of a free-space Green function is generally used. For the three-dimensional electrostatic problem

$$G(\bar{r}|\bar{r}') = \frac{1}{4\pi|\bar{r}-\bar{r}'|} \tag{1}$$

relates a unit source at $\bar{r}'$ to its effect at $\bar{r}$ with zero reference at infinity. For a source distribution $\rho(\bar{r}')$, the scalar potential field is given by the volume integration (Jaswon and Symm, (1977)

$$\phi(\bar{r}) = \frac{1}{\varepsilon} \int_V G(\bar{r}|\bar{r}')\rho(\bar{r}')dv' \tag{2}$$

where ε is the permitttivity of the medium.

Note that the Green function implicitly assumes that the intervening region is linear and homogeneous. Because partial differential equation methods relate reponses at immediately adjacent points, it is possible to use them – in the finite element method – for the solution of inhomogeneous, orthotropic (Wexler, 1977) and nonlinear media. On the other hand, because the boundary condition at infinity is embodied in the Green function, integral equation methods can cater for open-region problems.

Equation (2) describes a scalar IE problem. Vector field problems are similarly handled (e.g. Tai, 1971). For example, the law of Biot-Savart

$$\bar{B}(\bar{r}) = \mu \int_V \frac{\bar{J}(\bar{r}')\times(\bar{r}-\bar{r}')}{4\pi|\bar{r}-\bar{r}'|^3} dv' \tag{3}$$

can be written as

$$\bar{B}(\bar{r}) = \mu \int_V \bar{\bar{G}}(\bar{r}|\bar{r}')\cdot\bar{J}(\bar{r}')dv' \tag{4}$$

where the dyadic Green function is

66

$$\bar{\bar{G}}(\bar{r}|\bar{r}') = \frac{1}{4\pi|\bar{r}-\bar{r}'|^3}\left((r_x-r_x')(\hat{j}\hat{k}-\hat{k}\hat{j})\right) \tag{5}$$

$$+(r_y-r_y')(\hat{k}\hat{i}-\hat{i}\hat{k})+(r_z-r_z')(\hat{i}\hat{j}-\hat{j}\hat{i}))$$

and the electrical current density is

$$\bar{J}(\bar{r}') = \hat{i}J_x(\bar{r}')+\hat{j}J_y(\bar{r}')+\hat{k}J_z(\bar{r}') \tag{6}$$

Although the operations described by Equation (4) are more complicated than those in Equation (2), the form is virtually identical. And so, discussion of the scalar IE formulation suffices.

Rarely is a problem posed with a source distribution given and the potential distribution to be found. More frequently, a boundary condition is stated for a given impermeable surface. In this latter case, the problem is addressed by replacing the surface with an unknown source distribution. This distribution is then determined – using free-space analysis – in order that the required boundary condition be satisfied. When this is accomplished, the impermeable medium may be replaced without affecting the resulting fields. This is the simple layer approach. The significant advantage of the IE formulation that a three-dimensional problem is reduced to solving for a two-dimensional distribution over a surface. Similarly, a two-dimensional field problem is reduced to finding a distribution over a curve, i.e. it is reduced to a one-dimensional field problem.

Another formulation is obtained from Green's identity. Consider a source-free region in which the Laplace equation

$$\nabla^2\phi(\bar{r}) = 0 \tag{7}$$

holds. (A free source term may easily be accomodated in what follows.) Because the Green function is the solution of a point source,

$$-\nabla^2 G(\bar{r}\,|\,\bar{r}') = \delta(\bar{r}-\bar{r}') \tag{8}$$

where $\delta(\bar{r}-\bar{r}')$ is the Dirac delta function. Multiplying Equation (7) by $G(\bar{r}|\bar{r}')$ and (8) by $\phi(\bar{r}')$, substracting them and then integrating over a closed region R,

$$\phi(\bar{r}) = \int_R \left(G(\bar{r}|\bar{r}')\nabla^2\phi(\bar{r}')-\phi(\bar{r}')\nabla^2 G(\bar{r}|\bar{r}')\right)dv' \tag{9}$$

results. Employing Green's identity in order to transform the integration over R to an integration over a closed surface S,

$$\phi(\bar{r}) = \gamma\oint_S \left(G(\bar{r}|\bar{s}')\frac{\partial\phi(\bar{s}')}{\partial n'} - \phi(\bar{s}')\frac{\partial G(\bar{r}|\bar{s}')}{\partial n'}\right)ds' \tag{10}$$

67

where $\gamma = 0$ when $\bar{r}$ is not in R and $\gamma = 1$ when $\bar{r}$ is within R. Because $G(\bar{r}|\bar{s}')$ is singular when $\bar{r} = \bar{s}'$, careful analysis (e.g. Stakgold, 1968) shows that $\gamma = 2$ on the smooth part of S. For open-region problems, the contribution to the integral along the outer boundary vanishes and one need perform integrations only over surfaces of media discontinuities with S.

If a surface voltage distribution $\phi(s) = g(s)$ (i.e. a Dirichlet boundary condition) is given then, by rearranging Equation (10),

$$\oint_S G(\bar{s}|\bar{s}') \, \frac{\partial \phi(\bar{s}')}{\partial n'} \, ds' = b(s) \qquad (11)$$

results. It is a Fredholm integral equation of the first kind. The known voltage boundary condition is included within b(s) and the only unknown is $\frac{\partial \phi(\bar{s}')}{\partial n'}$.

Should a normal derivative (i.e. Neumann) boundary condition $\frac{\partial \phi(\bar{s})}{\partial n}$ be prescribed, then rearrangement of Equation (10) results in

$$\frac{1}{2} \, \phi(\bar{s}) + \oint_S \phi(\bar{s}') \, \frac{\partial G(\bar{s}|\bar{s}')}{\partial n'} ds' = b(s) \qquad (12)$$

which is a Fredholm integral equation of the second kind.

In multi-media problems, both the surface potential and the normal derivative at the surface are unknown. In such cases both $\phi(\bar{s})$ and $\frac{\partial \phi(\bar{s})}{\partial n}$ may be treated as unknowns in the resulting solution matrix formulation. Alternatively, by enforcing continuity of both the value and the normal derivative of potential across permeable interfaces, and by appropriate elimination of the normal derivative, one can obtain the integral equation

$$\frac{\varepsilon_{r+1}}{2} \, \phi(\bar{s}) + (\varepsilon_r - 1) \oint_S \phi(\bar{s}') \frac{\partial G(\bar{s}|\bar{s}')}{\partial n'} ds' = f(s) \qquad (13)$$

This involves only the surface potential as unknown but then requires solution of Equation (11) in order to obtain the normal derivative of the potential. Once this is accomplished the potential at any other point $\phi(\bar{r})$ may be obtained by employing Equation (10).

Simple-layer source distributions may also be used in multi-media problems (e.g. Klimpke, 1983) in a manner that retains a block-sparse structure as does Equation (10) for such cases.

68

3. SURFACE MODELLING

The traditional method of numerical solution, of integral equation problems, has been to approximate unknown source distributions as a set of pulses and to enforce the integral equation at one point within each pulse. For far-field computations, in regions that include planar objects, this method of pulse expansion and point matching yields acceptable results. However, for near-field computations − particularly near curved surfaces − the results are either unacceptable or are extremely costly to obtain. As a result, several workers explored higher-order surface and source modelling techniques.

Cruse (1974) employed linear variation of sources, over a surface modelled in a pulse fashion, and derived the discretized system of equations by point matching. Lachat and Watson (1976) described a parametric method for modelling both the source and the geometry. Discretization was accomplished by a point-matching technique. Replicating the isoparametric finite element technique Jeng and Wexler (1976,1977) described a method for boundary integral equation solution by modelling the surface in a finite-element sense using elements of various orders. Discretization was achieved by a variational/Galerkin approach.

Lagrangian elements
Borrowing from the isoparametric finite element method, one can map from local η−ξ coordinates to the global coordinates x-y-z by expressing each global coordinate as a polynomial function of the local coordinates. In the finite element sense, the polynomic variation over each such boundary element is expressed in terms of node points on the surface and Lagrangian finite element shape functions. Thus

$$x = \alpha^T(\xi,\eta)\mathbf{x} \ , \ y = \alpha^T(\xi,\eta)\mathbf{y} \ , \ z = \alpha^T(\xi,\eta)\mathbf{z} \qquad (14)$$

where the order of the vectors is determined by the element order. $\mathbf{x}$, $\mathbf{y}$ and $\mathbf{z}$ contain the coordinates of each node point in three-dimensional space and $\alpha(\xi,\eta)$ contains the associated shape functions. Figure 1 depicts a typical transformation.

By sharing edge nodes of adjacent elements, continuity of the surface (as in Figure 2) can be assured. However, tangents to the surfaces are discontinuous across element edges. This is therefore the cause of artificially high field intensities that have no physical reason for their existence. The cause is, of course, the creases produced by the modelling technique. It was found (Lean, 1981), however, that adjustment of the boundary element node locations could somewhat ameliorate this problem. Nonetheless, accuracy and economy were greatly improved compared with results obtained using lower-order methods. Results for two-dimensional, permeable-media problems are given elsewhere (Lean and Wexler, 1982).

Using appropriate Gauss quadrature formulas and geometric transformations of the region of integration, singular kernels may be handled numerically over arbitrarily curved surfaces (Lean and Wexler).

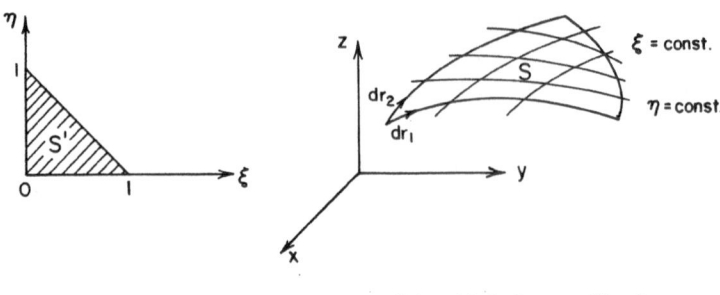

(a) Local coordinates   (b) Global coordinates

Figure 1.(a) The simplex element and (b) its transformation to three-space.   From Jeng and Wexler (1977).

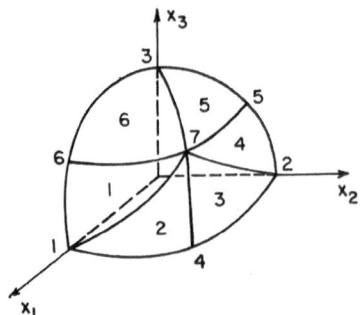

Figure 2. The boundary element model of a spheroid octant.
From Jeng and Wexler (1977)

Cubic spline elements

As mentioned in the Introduction, cubic splines permit the modelling of surfaces with second derivative continuity (e.g. Barsky and Greenberg, 1980). The B-spline basis function is defined by

70

$$B_i(x)=\begin{cases}0 & x < x_{i-2}\\(x-x_{i-2})^3/6h^3 & x_{i-2} < x < x_{i-1}\\(-3(x-x_{i-1})^3+3h(x-x_{i-1})^2+3h^2(x-x_{i-1})+h^3)/6h^3 & x_{i-1} < x < x_j\\(3(x-x_i)^3-6h(x-x_i)^2+4h^3)/6h^3 & x_i < x < x_{i+1}\\(-(x-x_i-1)^3+3h(x-x_{i-1})^2-3h^2(x-x_{i+1})+h^3)/6h^3 & x_{i+1} < x < x_{i+2}\\0 & x_{i+2} < x\end{cases} \quad (15)$$

and is depicted in Figure 3. This basis function is centred about the point x_i. To interpolate over an interval, a train of these basis functions is required. This is shown in Figure 4. Notice, that at most points, four B-splines must be involved in approximating a general function. At any one data point (e.g. x_{i-1}, x_i, x_{i+1}, etc.), three cubic splines interact. At a Lagrangian data point, only one shape function has a nonzero value and so Lagrangian interpolation is achieved simply by setting one coefficient equal to the value of the function of that point. By contrast, B-spline interpolation requires the solution of a tridiagonal system of equations. Although this is an additional complication, the computational cost is very low and benefits are considerable.

By considering an interval of width h, in Figure 4, we can define spline shape functions (Figure 5) in a manner analogous to Lagrangian shape functions. (Details are available in Bilgen (1982)). Over the interval $0 < \xi < 1$, Equations (15) take the form

$$\alpha_1(\xi)= (-\xi^3 + 3\xi^2 - 3\xi + 1)/6$$

$$\alpha_2(\xi) = (3\xi^3 - 6\xi^2 + 4)/6$$

$$\alpha_3(\xi) = (-3\xi^3 + 3\xi^2 + 3\xi + 1)/6 \quad (16)$$

$$\alpha_4(\xi) = \xi^3/6$$

The summation

$$P(\xi) = \sum_{j=1}^{4} \alpha_j(\xi) V_{i+j-2} \quad (17)$$

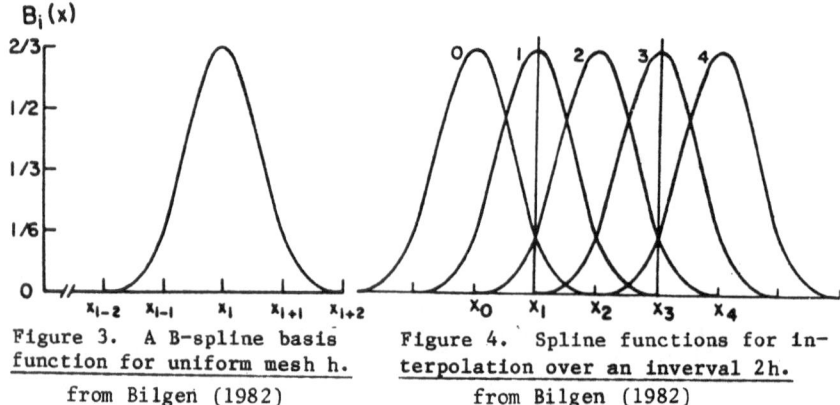

Figure 3. A B-spline basis
function for uniform mesh h.

from Bilgen (1982)

Figure 4. Spline functions for interpolation over an inverval 2h.

from Bilgen (1982)

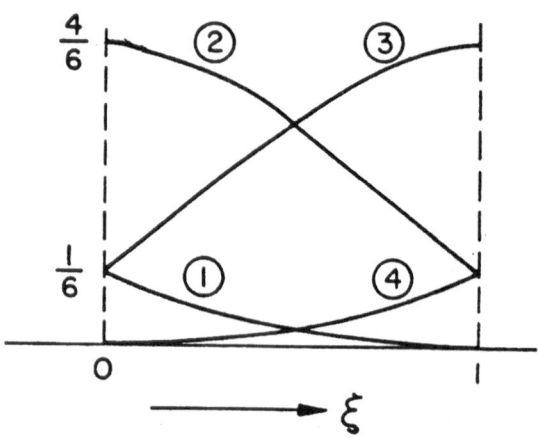

Figure 5. Element cubic spline shape functions.
From Bilgen and Wexler (1982).

defines a mapping from ξ to x, y or z or an interpolation of a potential function as in the Lagrangian scheme. In this spline case, however, the coefficients do not individually correspond to values of
the interpolated function.

A surface in three-dimensional space is expressed by employing the
mapping (17) to define four intersecting space curves as shown in
Figure 6. Two of the curves are expressed in terms of the parameter
η and the other two are in terms of ξ. Without going into detail

72

here, the enclosed space is filled in by multiplying each space curve (say, FO(η)) by a spline-blending function of the other parameter (i.e. $\beta 0(\xi)$) thus producing a continuous function over the enclosed space. This is done for each space curve and the result is summed with a function subtracted to correct for the duplication at the corners. The resulting surface is known as a Coons patch.

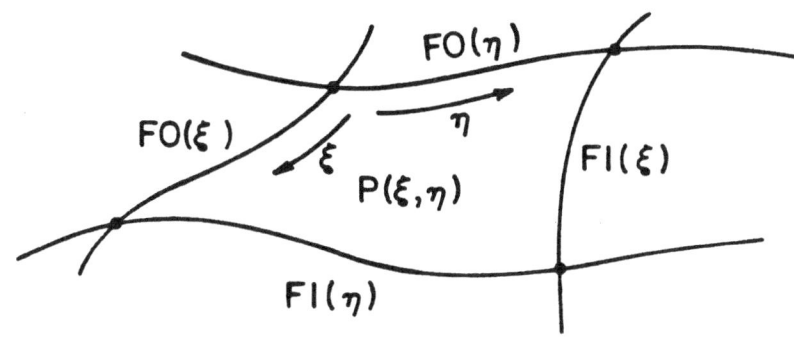

Figure 6. A Coons patch
From Bilgen (1982)

Because of the high-order surface modelling and because all nodes are shared with adjacent elements (thus reducing storage requirements), high accuracy at low computational costs are attainable. These goals have not yet been fully attained, but work is proceeding in this direction. Possible schemes involve special techniques for block-sparse systems and diakoptical techniques (Wexler, 1980).

4. A NEW BEM PROGRAM

Many of the ideas, discussed in this paper, are being implemented in a new spline, boundary element method code.

The program solves three-dimensional fields in a region that may consist of piecewise-homogeneous media. Conducting surfaces at unknown potentials may be accommodated. Surfaces are modelled using spline techniques in order to eliminate accidental creases and a Galerkin method is used in order to ensure stability. The program caters for automatic numerical integration of the Green function singularity over arbitrarily-curved surfaces. It also embodies an automated mesh generator based on the spline modelling technique (Yildir and Wexler, 1982). Preliminary results indicate high accuracy is attainable for both scalar and vector fields.

73

5. REFERENCES

Barsky, B.A. and Greenberg, D.P., "Determining a set of B-spline control vertices to generate an interpolating surface," Computer Graphics and Image Processing, Vol. 14, pp. 203-226 (1980).

Bilgen, S. and Wexler, A., "Spline boundary element solution of dielectric scattering problems ," Proceedings, 12th European Microwave Conference, Microwave Exhibitions and Publishers Ltd., Tunbridge Wells, England, pp. 371-377 (1982).

Bilgen, S., Cubic Spline Elements for Boundary Integral Equations. Ph.D. dissertation, Report TR82-5, Department of Electrical Engineering, University of Manitoba, Winnipeg (1982).

Cruse, T.A., "An improved boundary-integral equation method for three-dimensional elastic stress analysis," Computers and Struct., Vol. 4, pp. 741-754 (1974).

Jaswon, M.A. and Symm, G.T., Integral Equation Methods in Potential Theory and Elastostatics. Academic Press, London (1977).

Jeng, G. and Wexler, A., "Finite-element solution of boundary integral equations," in Proc. Int. Symp. Large Engng Systems (Ed. A. Wexler), Pergamon Press, Oxford, pp. 112-121 (1977).

Jeng, G. and Wexler, A., "Isoparametric, finite element, variational solution of integral equations for three-dimensional fields," Int. J. Num. Meth. Engng, Vol. 11, pp. 1455-1471 (1977).

Klimpke, B.W., A Two-Dimensional, Multi-Media Boundary Element Method. M.Sc. dissertation, Report TR83-2, Department of Electrical Engineering, University of Manitoba, Winnipeg. (1983).

Lachat, J.C. and Watson, J.O., "Effective numerical treatment of boundary integral equations: a formulation for three-dimensional elastostatics," Int. J. Num. Meth. Engng , Vol. 10, pp. 991-1005 (1976).

Lean, M.H., Electromagnetic Field Solution with the Boundary Element Method. Ph.D. dissertation, Report TR81-5, Department of Electrical Engineering, University of Manitoba, Winnipeg (1981).

Lean, M.H. and Wexler, A., "Accurate field computation with the boundary element method", IEEE Trans. Magnetics, Vol. MASG18, No. 2 pp. 332-335 (1982).

Lean, M.H. and Wexler, A., "Accurate numerical integration of singular boundary element kernels over boundries with curvature." Submitted for publication.

Stakgold, I., Boundary Value Problems of Mathematical Physics. Vol. II, Macmillan, New York (1968).

Tai, C.-T., Dyadic Green's Functions in Electromagnetic Theory. Intext, Scranton, Pennsylvania.

Wexler, A., "Isoparametric finite elements for continuously inhomogeneous and orthotropic media," in Finite Elements in Water Resources (Eds. W.G. Gray, G.F. Pinder, and C.A. Brebbia), Pentech Press, London, pp. 2.3-2.24 (1976).

Wexler, A., Perspectives on the Solution of Simultaneous Equations, Report TR79-2, Department of Electrical Engineering, University of Manitoba, Winnipeg (1980).

Yildir, Y.B., and Wexler, A., MANDAP A FEM/BEM Data Prepartion Package - User's Manual. Report TR82-3, Department of Electrical Engineering, University of Manitoba, Winnipeg (1982).

MAGNETOSTATIC FIELD CALCULATIONS ASSOCIATED WITH
THICK SOLENOIDS WITH IRON PRESENT

J Caldwell[1], A Zisserman[2] and R Saunders[2]

[1] School of Mathematics, Statistics and Computing, Newcastle-upon-Tyne Polytechnic

[2] Department of Mathematics and Computer Studies, Sunderland Polytechnic

1. INTRODUCTION

The magnetostatic field associated with iron-free axisymmetric systems has been considered in detail by Boom and Livingston (1962), Garrett (1951) and many others. More recently Caldwell (1982), Caldwell and Zisserman (1983a, 1983b) have carried out some work which takes account of the effects of the presence of iron on such systems. The two main advantages of introducing iron are:

i) higher fields are provided for the same current resulting in substantial savings for conventional conductors,

ii) field uniformity is improved even for superconducting solenoids by placing the iron in a suitable position.

The geometry to be considered is shown in Figure 1. A toroidal conductor region V' of rectangular cross-section is located midway between two semi-infinite blocks of iron. The remaining region V between the conductor and iron is assumed to be insulating.

Using cylindrical polar coordinates (r, ϕ, z) Caldwell (1982) has shown that the field equations can be reduced to the differential system

$$\nabla_1^2 \, A_\phi = \begin{array}{ll} 0 & \text{in} \ \ V \\ -c & \text{in} \ \ V' \end{array} \qquad (1)$$

$$\frac{\partial A_\phi}{\partial z} = 0 \quad \text{on} \ \ z = 0,1 \qquad (2)$$

$$A_\phi = 0 \quad \text{on} \ \ r = 0 \qquad (3)$$

$$A_\phi \to 0 \quad \text{as} \ \ r \to \infty \qquad (4)$$

where

$$\nabla_1^2 = \frac{\partial^2}{\partial r^2} + \frac{1}{r} \frac{\partial}{\partial r} - \frac{1}{r^2} + \frac{\partial^2}{\partial z^2} \qquad (5)$$

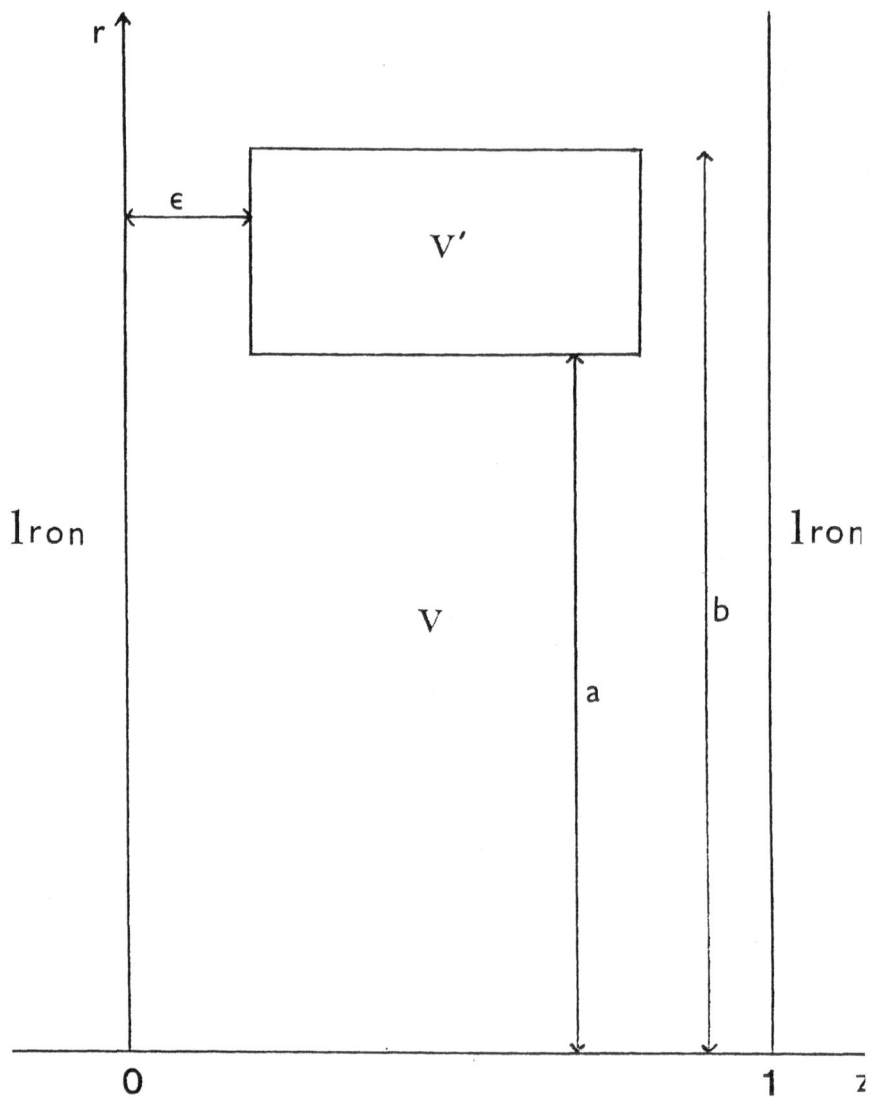

Figure 1. Geometry of the system

The vector potential $\underline{A}$ is given by $\underline{A} = A_\phi \underline{1}_\phi$ and the conductor current density j by $j = c/\mu_0$ where μ_0 is the permeability of free space.

The solution for the simple case $\varepsilon = 0$ is discussed in Section 2 using the integral representation of the vector potential $\underline{A}$. This work is extended in Section 3 using a perturbation analysis with ε as a small parameter. In this way a simple analytic expression is obtained which describes the field close to the axis of symmetry. For later comparison purposes a brief account is given in Section 4 of the Fourier series solution of Caldwell and Zisserman (1983b). In order to solve the system exactly it was necessary in all of this work to make the assumption that the iron had infinite permeability. For practical situations, where with superconducting magnets there might well be some saturation, it is important to know the implications of this assumption.

The integral representation of the vector potential $\underline{A}$ and the method of images is extended in Section 5 to obtain a series expansion of the vector potential for iron of constant finite permeability μ. In Section 6 comparison is made between the results and those obtained by the exact solution for the extreme cases of infinite permeability and unit permeability (i.e. no iron). The advantages of this method are also discussed.

2. SOLUTION FOR THE CASE $\varepsilon = 0$.

The solution of the differential system for the simple case $\varepsilon = 0$ can easily be found. The vector potential due to the current distribution may be obtained from the integral representation

$$\underline{A}(\underline{r}) = \frac{\mu_0}{4\pi} \int_{V'} \frac{\underline{j}(\underline{r}')}{|\underline{r}-\underline{r}'|} \, dV' \tag{6}$$

This gives

$$A_\phi = \frac{\mu_0 j}{4\pi} \int_a^b \int_0^{2\pi} \int_{-\infty}^{\infty} \frac{x \cos\theta \, dx \, d\theta \, dz'}{\{(z-z')^2+r^2+x^2-2xr\cos\theta\}^{\frac{1}{2}}}$$

$$= \frac{\mu_0 j}{2\pi} \int_a^b \int_0^{2\pi} x \cos\theta \, dx \, d\theta \lim_{R\to\infty} \int_0^R \frac{dZ}{(Z^2+p^2)^{\frac{1}{2}}}$$

(on setting $Z = z - z'$, $p^2 = r^2 + x^2 - 2xr \cos\theta$)

79

$$= \frac{\mu_0 j}{2\pi} \lim_{R \to \infty} \int_a^b \int_0^{2\pi} x \cos\theta \sinh^{-1}(R/p) \, dx \, d\theta$$

Now as $x \to \infty$, $\sinh^{-1} \sim \ln 2x + \dfrac{1}{4x^2} + \dots$

This means that

$$A_\phi = \frac{\mu_0 j}{2\pi} \lim_{R \to \infty} \int_a^b x \, dx \int_0^{2\pi} \cos\theta \, (\ln 2R - \ln p) d\theta$$

$$= \frac{\mu_0 j}{4\pi} \int_a^b x \, dx \int_0^{2\pi} \cos\theta \, \ln (r^2 + x^2 - 2xr \cos\theta) d\theta$$

$$= \frac{\mu_0 j r}{2\pi} \int_a^b x^2 dx \int_0^{2\pi} \frac{\sin^2\theta \, d\theta}{r^2 + x^2 - 2xr\cos\theta}$$

(on integrating by parts)

$$= \frac{\mu_0 j r}{4\pi} \int_a^b dx \int_0^{2\pi} \frac{(1-\cos2\theta)d\theta}{(r/x)^2 + 1 - 2(r/x)\cos\theta}$$

$$= \frac{\mu_0 j r}{2} \int_a^b \left\{ \frac{1}{1-(r/x)^2} - \frac{(r/x)^2}{1-(r/x)^2} \right\} dx$$

$$= \mu_0 j (b-a) r/2 \qquad\qquad\qquad (7)$$

80

3. PERTURBATION ANALYSIS (INFINITE μ)

The potential for the system with a small iron air gap of width ε can be derived by perturbing the $\varepsilon = 0$ case. The expansion is derived from the integral representation of the vector potential $\underline{A}$ given in Equation (6). The potential is obtained by writing

$$A_\phi = \psi_0 + \Psi \tag{8}$$

where ψ_0 is the $\varepsilon = 0$ case described in Section 2 and Ψ is the potential due to a set of image coils which represent the gap between the conductor and the iron.

We already know that

$$\psi_0 = \mu_0 j(b-a)r/2 \qquad \text{for} \quad r < a$$

Hence

$$A_\phi = \psi_0 - \frac{\mu_0 j}{4} \sum_{n=-\infty}^{\infty} \int_a^b \int_0^{2\pi} \int_{-\varepsilon}^{\varepsilon} \frac{x \cos\theta \, dx \, d\theta \, dz'}{\{(z-z'-n)^2 + r^2 + x^2 - 2xr\cos\theta\}^{\frac{1}{2}}} \tag{9}$$

Caldwell (1982) shows that this results in a simple analytic expression which describes the field close to the axis of symmetry, namely

$$A_\phi = \frac{\mu_0 j(b-a)r}{2} - \frac{\mu_0 j \varepsilon r}{2} \{F(b,z) - F(a,z)\} + 0(\varepsilon^3) \tag{10}$$

where

$$F(c,z) = \sum_{n=-\infty}^{\infty} \{\tfrac{1}{2}\ell n \left(\frac{1+\sin\phi}{1-\sin\phi}\right) - \sin\phi\} \tag{11}$$

and

$$\sin\phi = c\{(z-n)^2 + c^2\}^{-\frac{1}{2}} \tag{12}$$

The magnetic induction $\underline{B}$ is obtained from $\underline{B} = \text{curl}\underline{A}$, that is

$$B_r = -\frac{\partial A_\phi}{\partial z} \quad , \quad B_z = \frac{1}{r}\frac{\partial}{\partial r}(rA_\phi) \tag{13}$$

For example,

$$B_z = \mu_0 j(b-a) - \mu_0 j \varepsilon\{F(b,z) - F(a,z)\} \tag{14}$$

In the neighbourhood of the axis of symmetry the field B_z is therefore obtained by substituting the expression for $F(c,z)$ in Equation (11) into this Equation (14). It should be noted that the above analysis applies for infinite permeability μ. The finite permeability case will be discussed in Section 5.

81

4. FOURIER SERIES APPROACH

The differential system posed by Equations (1)-(5) may be solved by examining Fourier type solutions. We divide the domain into the three regions $r \leqslant a$, $a \leqslant r \leqslant b$, $r \geqslant b$.

For $r \leqslant a$ and $r \geqslant b$

$$\nabla_1^2 A_\phi = 0$$

subject to the conditions

$$A_\phi = 0 \quad \text{on} \quad r = 0$$
$$A_\phi \to 0 \quad \text{as} \quad r \to \infty$$

We can solve directly by separation of variables.

For $a \leqslant r \leqslant b$,

$$\nabla_1^2 A_\phi = \begin{cases} -c & \varepsilon \leqslant z \leqslant 1-\varepsilon \\ 0 & \text{otherwise} \end{cases}$$

subject to the conditions

$$\frac{\partial A_\phi}{\partial z} = 0 \quad \text{on} \quad z = 0,1.$$

To solve the homogeneous form we separate the variables and to find the particular integral we look for a Fourier cosine type solution, namely

$$A_\phi(r,z) = f_0(r) + \sum_{n=1}^{\infty} f_n(r) \cos n\pi z \tag{15}$$

The solutions are matched up by using the continuity conditions of A_ϕ and $\frac{\partial A_\phi}{\partial r}$ at the boundaries $r = a$ and $r = b$.

The most important region for considering field uniformity is $r \leqslant a$ and Caldwell and Zisserman (1983b) have produced the solution

$$A_\phi = \frac{\mu_0 j(1-2\varepsilon)(b-a)r}{2} + \sum_{\substack{(n \text{ even} \\ \text{and} \neq 0)}} M_n U_n I_1 (n\pi r) \cos n\pi z \tag{16}$$

where

$$M_n = \frac{4(-1)^{n/2}\mu_0 j}{(n\pi)^3} \sin \frac{n\pi}{2} (1-2\varepsilon), \tag{17}$$

$$U_n = \int_{n\pi a}^{n\pi b} t\, K_1(t)\, dt, \qquad (18)$$

and I_1, K_1 are the modified Bessel functions of the first and second kind.

This series for A_ϕ converges very quickly as can be seen by considering the results in Table 1 for the geometry $a = 0.9$, $b = 1.1$, $\varepsilon = 0.05$, $c = \mu_0 j = 100$. The behaviour of $A_\phi(r,z)$ is dominated by the first (n=0) term, namely, $\mu_0 j(1-2\varepsilon)(b-a)r/2$ which equals 0.9, and Table 1 is typical of the behaviour shown throughout the region.

n	$M_n\, U_n\, I_1\ (n\pi r)\ \cos n\pi x$ $r=0.1,\ z = 0.5$
2	1.354×10^{-3}
4	-4.556×10^{-6}
6	1.563×10^{-8}
8	-5.628×10^{-11}
10	2.087×10^{-13}

Table 1.   Values of $M_n U_n I_1$ $(n\pi r)$ $\cos n\pi z$ where $r = 0.1$, $z = 0.5$ for $n = 2(2)10$.

The radial and axial components of $\underline{B}$ can be found from Equations (13) and (16) and are

$$B_r(r,z) = \sum_{\substack{(n\ \text{even} \\ \text{and}\ \neq 0)}} n\pi\, M_n\, U_n\, I_1\ (n\pi r)\ \sin n\pi z \qquad (19)$$

and

$$B_z(r,z) = \mu_0 j(1-2\varepsilon)(b-a) + \sum_{\substack{(n\ \text{even} \\ \text{and}\ \neq 0)}} n\pi\, M_n\, U_n\, I_0\ (n\pi r)\ \cos n\pi z \qquad (20)$$

In B_z the behaviour is dominated by the n=0 term which is now constant. In B_r, since the large n=0 term is removed by the differentiation, we find that $|B_r| \ll |B_z|$ and this is exactly the requirement for a uniform field.

83

5.   SERIES EXPANSION (FINITE μ)

For finite μ the effect of the iron is built in by including an infinite set of image coils in which case Equation (6) in Section 2 becomes

$$A_\phi(r,z) = \frac{\mu_0 j}{4\pi} \sum_{n=-\infty}^{\infty} \int_a^b \int_0^{2\pi} \int_\epsilon^{1-\epsilon} \frac{k^{|n|} x \cos\theta \; dx \; d\theta \; dz'}{\{(z-z'-n)^2 + r^2 + x^2 - 2xr\cos\theta\}^{\frac{1}{2}}} \tag{21}$$

where $k = (\mu-1)/(\mu+1)$ is the image factor and μ is the permeability of the iron. In this case of finite μ a perturbation analysis using ε as a small parameter is not possible. However, we can expand the potential A_ϕ using a Maclaurin series expansion in r, namely

$$A_\phi(r,z) = A_\phi(0,z) + r \frac{\partial A_\phi}{\partial r}(0,z) + \frac{r^2}{2!} \frac{\partial^2 A_\phi}{\partial r^2}(0,z) + O(r^3) \tag{22}$$

Writing

$$I_n = \int_0^{2\pi} \frac{\cos\theta \; d\theta}{R}$$

where

$$R = \{(z-z'-n)^2 + x^2 + r^2 - 2xr\cos\theta\}^{\frac{1}{2}}$$

gives

$$A_\phi(r,z) = \frac{\mu_0 j}{4\pi} \sum_{n=-\infty}^{\infty} k^{|n|} \int_a^b \int_\epsilon^{1-\epsilon} x \, I_n \, dx \, dz'$$

This means that the Maclaurin series expansion (22) is

$$A_\phi(r,z) = \frac{\mu_0 j}{4\pi} \sum_{n=-\infty}^{\infty} k^{|n|} \int_a^b x \, dx \int_\epsilon^{1-\epsilon} \{I_n\big|_{r=0} + r \frac{\partial I_n}{\partial r}\big|_{r=0}$$

$$+ \frac{r^2}{2!} \frac{\partial^2 I_n}{\partial r^2}\big|_{r=0} + O(r^3)\}dz' \tag{23}$$

If we note that

$$\int_0^{2\pi} \cos^{2n+1}\theta \; d\theta = 0$$

84

then

$$I_n\big|_{r=0} = 0$$

$$\frac{\partial I_n}{\partial r}\bigg|_{r=0} = \frac{\pi x}{(w^2+x^2)^{3/2}} \quad \text{where } w = (z-z'-n)$$

$$\frac{\partial^2 I_n}{\partial r^2}\bigg|_{r=0} = 0$$

Thus, after carrying out the integrations with respect to x and z', Equation (23) becomes

$$A_\phi(r,z) = \frac{\mu_0 j r}{4} \sum_{n=-\infty}^{\infty} k^{|n|} [[w \ln\{x + (x^2 + w^2)^{\frac{1}{2}}\}]_{x=a}^{x=b}]_{w=\epsilon+n-z}^{w=1-\epsilon+n-z}$$

$$(24)$$

The radial and axial components of the magnetic induction are

$$B_r(r,z) = \frac{\mu_0 j r}{4} \sum_{n=-\infty}^{\infty} k^{|n|} [[\ln\{x+(x^2+w^2)^{\frac{1}{2}}\}-x(w^2+x^2)^{-\frac{1}{2}}]_{x=a}^{x=b}]_{w=\epsilon+n-z}^{w=1-\epsilon+n-z}$$

$$+ 0(r^3) \qquad (25)$$

and

$$B_z(r,z) = \frac{\mu_0 j}{2} \sum_{n=-\infty}^{\infty} k^{|n|} [[w\ln\{x + (x^2 + w^2)^{\frac{1}{2}}\}]_{x=a}^{x=b}]_{w=\epsilon+n-z}^{w=1-\epsilon+n-z}+0(r^2)$$

$$(26)$$

6. DISCUSSION OF RESULTS

The geometry considered has parameters

$$a = 0.9, \ b = 1.1, \ \epsilon = 0.05, \ c = \mu_0 j = 100.$$

In Tables 2, 3 and 4 values of $A_\phi(r,z)$, $B_r(r,z)$ and $B_z(r,z)$ are obtained at points close to the axis ($r = 0.1, 0.2, 0.3, 0.4, 0.5$ with $z = 0.1$; $z = 0.2, 0.3, 0.4, 0.5$ with $r = 0.1$) for a range of values of permeability ($\mu = 1, 10, 10^2, 10^3, \infty$). The results obtained from the series expansion method in Section 5 are compared with the exact Fourier series results (for $\mu = \infty$) and the Legendre Polynomial results of Garrett (for $\mu = 1$). All the results in the tables are calculated by summing the series from $N = -100$ to $N = +100$. It is interesting to note that doubling the number of terms included did not affect the values to 4 significant figures.

85

r	z	F.S.	Series Expansion					L.P.
			μ=∞	μ=10³	μ=10²	μ=10	μ=1	
0	0.1	0	0	0	0	0	0	0
0.1	0.1	0.8989	0.8989	0.8971	0.8812	0.7579	0.3491	0.3496
0.2	0.1	1.797	1.798	1.794	1.762	1.516	0.6982	0.7023
0.3	0.1	2.695	2.698	2.691	2.644	2.274	1.047	1.061
0.4	0.1	3.592	3.596	3.588	3.525	3.032	1.400	1.429
0.5	0.1	4.485	4.495	4.485	4.406	3.790	1.746	1.811
0.1	0.2	0.8996	0.8996	0.8978	0.8825	0.7645	0.3747	0.3754
0.1	0.3	0.9004	0.9004	0.8986	0.8838	0.7698	0.3954	0.3954
0.1	0.4	0.9011	0.9010	0.8993	0.8847	0.7732	0.4069	0.4080
0.1	0.5	0.9013	0.9012	0.8996	0.8851	0.7743	0.4112	0.4123

Table 2.   Comparison of values of $A_\phi(r,z)$

As can be seen from Table 2 there is excellent agreement between the series expansion of A_ϕ and the known infinite and unit permeability limits. At r = 0.1 there is largely agreement to 3 significant figures and as far out as r = 0.4 the agreement is still satisfactory. On moving from the infinite case to $\mu = 10^2$ there is only a 1% change in the values although they have more than halved by $\mu = 1$. Thus for reasonably high permeabilities (say $\mu > 50$) the assumption of infinite permeability is a good model of the situation.

r	z	F.S.	Series Expansion					L.P.
			μ=∞	μ=10³	μ=10²	μ=10	μ=1	
0.1	0.1	5.056 E-3	4.807 E-3	5.595 E-3	1.125 E-2	7.132 E-2	0.2792	0.2816
0.2	0.1	1.172 E-2	9.615 E-3	1.119 E-2	2.511 E-2	0.1426	0.5584	0.5781
0.3	0.1	2.222 E-2	1.442 E-2	1.679 E-2	3.766 E-2	0.2140	0.9376	0.9062
0.4	0.1	4.045 E-2	1.923 E-2	2.238 E-2	5.022 E-2	0.2853	1.117	1.287
0.5	0.1	7.411 E-2	2.404 E-2	2.798 E-2	6.277 E-2	0.3566	1.396	1.748
0.1	0.2	8.126 E-3	7.733 E-3	8.351 E-3	1.381 E-2	6.017 E-2	0.2300	0.2317
0.1	0.3	8.058 E-3	7.678 E-3	8.102 E-3	1.186 E-2	4.387 E-2	0.1633	0.1648
0.1	0.4	4.947 E-3	4.718 E-3	4.934 E-2	6.847 E-3	2.318 E-2	8.483 E-2	8.563 E-2
0.1	0.5	0	0	0	0	0	0	0

Table 3.   Comparison of values of $B_r(r,z)$

Similarly Table 3 shows that B_r gives good agreement at both limits. The effect of the iron in suppressing B_r so that the field lines are perpendicular at the iron surface can be clearly seen by the reduction of the component by several orders of magnitude compared to the $\mu = 1$ (iron-free) case.

r	z	F.S.	Series Expansion					L.P.
			$\mu=\infty$	$\mu=10^3$	$\mu=10^2$	$\mu=10$	$\mu=1$	
0	0.1	17.98	17.98	17.94	17.62	15.16	6.982	6.982
0.1	0.1	17.98	"	"	"	"	"	6.982
0.2	0.1	17.97	"	"	"	"	"	7.002
0.3	0.1	17.96	"	"	"	"	"	7.063
0.4	0.1	17.93	"	"	"	"	"	7.167
0.5	0.1	17.88	"	"	"	"	"	7.315
0.1	0.2	17.99	17.99	17.96	17.65	15.29	7.494	7.523
0.1	0.3	18.01	18.01	17.97	17.68	15.40	7.889	7.926
0.1	0.4	18.02	18.02	17.99	17.70	15.46	8.139	8.180
0.1	0.5	18.03	18.02	17.99	17.70	15.49	8.224	8.267

Table 4.   Comparison of values $B_z(r,z)$

Table 4 shows the B_z component. The effect of the iron both in producing a uniform field and in boosting the field is obvious. In the iron-free case the field changes by more than 15% between $z = 0.1$ and $z = 0.5$ at $r = 0.1$ whilst with iron of high permeability present it changes by less than 0.5% over the same range and the field is more than twice as large.

It should be noted that the B_z expression is accurate only to $O(r^2)$ whilst the others are accurate to $O(r^3)$. Of course, more information could be gained about the variation of B_z with r by including more terms in the Maclaurin series expansion.

7.   CONCLUSIONS

The Maclaurin series expansion method described in Section 5 can be easily computerised and provides a quick and flexible method for calculating the effects of iron of finite permeability μ on coil systems. The effects of the iron in boosting and improving the uniformity of the field are clearly demonstrated. The accuracy could be extended to a larger region by including more terms in the Maclaurin series expansion.

REFERENCES

1. Boom, R. W. and Livingston, R. S., "Superconducting solenoids",
    Proc. I.R.E., 274 (1962).

2. Caldwell, J., "Magnetostatic field calculations associated with
    superconducting coils in the presence of magnetic material",
    IEEE Transactions on Magnetics, Vol. MAG-18, 2, 397 (1982).

3. Caldwell, J. and Zisserman, A., "Magnetostatic field calculation
    in the presence of iron using a Green.'s function approach",
    J.Appl.Phys., 54, 2 (1983 a).

4. Caldwell, J. and Zisserman, A., "A Fourier series approach to
    magnetostatic field calculations involving magnetic material",
    accepted for publication in J.Appl.Phys. (1983 b).

5. Garrett, M. W., "Axially symmetric systems for generating and
    measuring magnetic fields", J.Appl.Phys., 22, 1091 (1951).

PERIODIC SOLUTIONS FOR CERTAIN NON-LINEAR

PARABOLIC PARTIAL DIFFERENTIAL EQUATIONS

G. Gregory

NEI Parsons Limited, Newcastle upon Tyne NE6 2YL

1. INTRODUCTION

A fast method of calculating periodic solutions to certain non-linear
parabolic partial differential equations has been developed for two
particular electromagnetic problems. The range of application could
be wider.

A typical situation for applying the method is where a region of non-
linear behaviour, often confined to a narrow skin, is surrounded by
a region of linear behaviour. The thinness of the non-linear region
means that it can be described in terms of one less space dimension,
but because of the non-linearity it gains the time dimension. The
surrounding linear region has no time dimension but retains the space
dimension that is lost to the thin region.

The method of solution is in four stages, and will be described in its
application to the problem of voltage distribution in stress relieving
tape, and then as applied to an eddy current problem.

2. VOLTAGES IN STRESS RELIEVING TAPE

Physical background
In a generator stator the copper conductor bars which carry current
at up to 25 kV rms are bound with insulating tape bearing mica in
epoxy resin. The bars lie in axial slots in the stator body, and
emerge from the slots at the core end to continue in the same
direction for a short distance, known as the straight standout, before
being wound around the cone of the stator (see Figure 1). The stator
core is at earth potential, and the conductor bars at high voltage,
and the electrical stresses would cause flash-over and breakdown if
there were no provision to relieve them.

A widely used remedy is to apply semi-conducting material to the
outside of the conductor bar insulation on the straight standout
section, so as to smooth the voltage gradient near the core end. This
stress relieving material can be in the form of either varnish or tape
containing silicon carbide granules, and the work described below was
to calculate the voltage distribution along the surface of such tape.

89

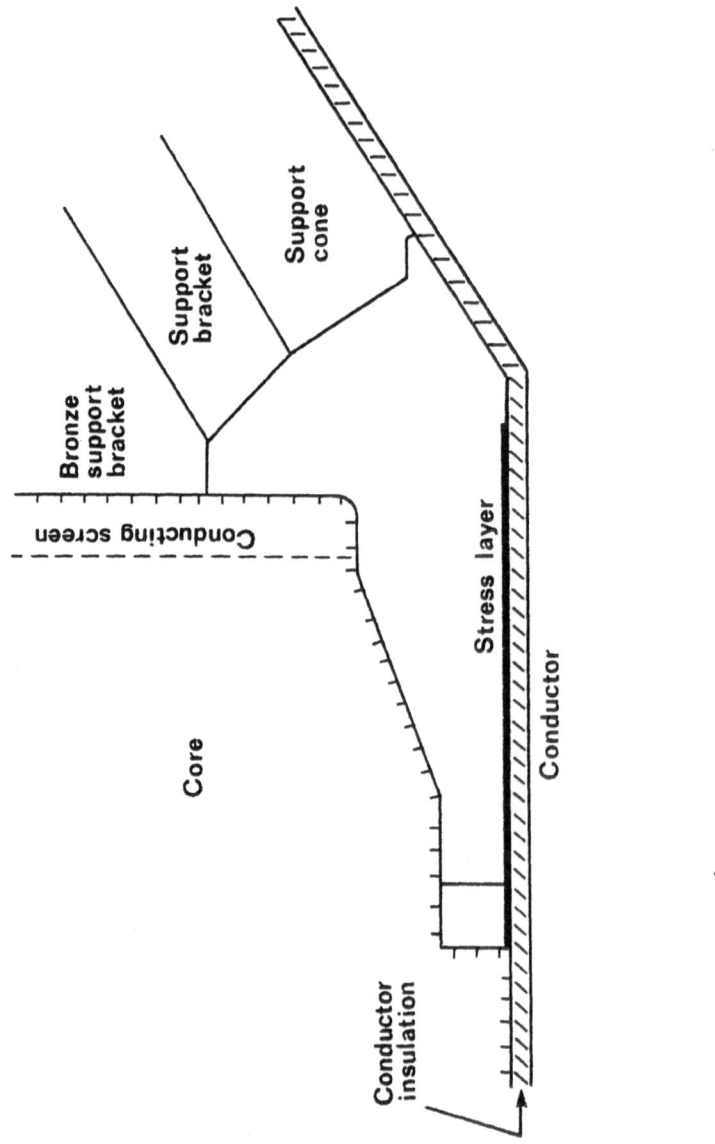

Figure 1.   2-D representation of end region

Method of solution

The parabolic equation to be solved in the tape is derived in the Appendix:

$$G\frac{\partial^3 V}{\partial x^2 \partial t} + H\frac{\partial}{\partial x}F\left(\frac{\partial V}{\partial x}\right) - \frac{\partial V}{\partial t} = H\frac{\partial D_n}{\partial t}, \tag{1}$$

where G, H and K are constants, V is voltage on the tape surface, D_n is normal electric displacement on the tape surface and F is the non-linear function relating current to electric force. The tape is subdivided along its length, the x direction, but is treated as being infinitesimally thin. The method of calculation can be broken down into four stages.

The first stage involves only the exterior linearly behaved region of insulation. By solving Laplace's equation the voltages at discrete points along the tape layer, V_m, can be linked to the normal electric displacements there, D_{nm}. A finite element method is used to build up the coefficient matrix M which expresses the linear relationship

$$\underline{V} = M\underline{D_n}, \tag{2}$$

which is independent of time. This stage, in effect, uses the finite element solution to establish Equation (2) as the boundary condition imposed by the exterior fields on the fields in the tape. After the matrix M has been determined, no further work is required on the exterior fields.

The second stage of solution deals with the interior non-linear problem, and sets up starting values for the voltages. Efficiency of solution of any non-linear problem is strongly dependent on the quality of its starting values, so this crucial stage will be explained in greater detail. The interior problem is discretised in two dimensions – distance x along the tape layer, and a range of points in time t. Working with the parabolic equation in discretised form, for each point x_m a numerical Fourier analysis in time is performed to divide the equation into the sine and cosine parts of the fundamental harmonic.

This gives the voltage at any point x_m along the tape in the form

$$\text{Voltage at } x_m = (V_m + jW_m), \tag{3}$$

where j is the imaginary unit, and the usual convention is followed for representing alternating fields by complex algebra. V_O and W_O are known. It is supposed that there is an exponential variation in the voltage at the next two points x_{m+1}, x_{m+2} such that

$$\text{Voltage at } x_{m+1} = (V_m + jW_m)\exp\{-(a_m + jb_m)(x_{m+1} - x_m)\}, \tag{4}$$

$$\text{Voltage at } x_{m+2} = (V_m + jW_m)\exp\{-2(a_m + jb_m)(x_{m+1} - x_m)\}.$$

a_m and b_m are found by equating to zero the sine and cosine parts of the fundamental at x_{m+1} of the discretised equation, and solving by a numerical Newton-Raphson iteration. Then the voltage at x_{m+1} can be written in the form of Equation (3) as

$$\text{Voltage at } x_{m+1} = (V_{m+1} + jW_{m+1})$$

and the process applied again for x_{m+2}. In this way a set of starting values V and W are built up along the tape, which define the voltage in terms of the fundamental harmonic.

The third stage continues in the fundamental harmonic, but brings in the matrix M from the first stage. A complete solution for the fundamental harmonic is obtained by a Newton-Raphson iteration from the starting values. The right hand sides change slowly through the effect of the exterior solution, but this spoils the convergence very little.

The fourth stage is to calculate the solution of the complete problem, bringing in as many time harmonics as may be required.

With starting values defined by the fundamental solution of stage three, a weighted norm of residuals is optimised for increments based on those which would appear in a relaxation procedure. After each Newton-Raphson step new increments are computed to be used in redefining the optimisation. When the residuals have been reduced sufficiently far, the full solution has been obtained.

Results
Calculations. A comparison was made between stages two, three and four above, and Figure 2 illustrates not only how well the fundamental harmonic solution approximates the final solution, but also the quality of the starting values. The calculation was made using a conductor bar voltage of 50 kV rms, discretised with 36 intervals along the tape; the final solution includes time harmonics up to the fifth. Further calculations showed that to include eleven harmonics made only about 0·6% difference to the final voltages, and that using a finer discretisation of 72 x values gave less than 1% change in the final voltages. All calculations were made for a conductor bar current frequency of 50 Hz.

The next step was to investigate the change of voltage distribution with change in tape characteristics. The Figure 3 shows that the distribution is in effect drawn nearer to the core end for tape of lower conductivity. Experimental investigations have shown that tape characteristics can vary considerably under different conditions. All experimental work to which the author makes reference is contained in Wood (1982). One factor involved can be the grain size of the silicon carbide granules, which may not be uniform, and some

92

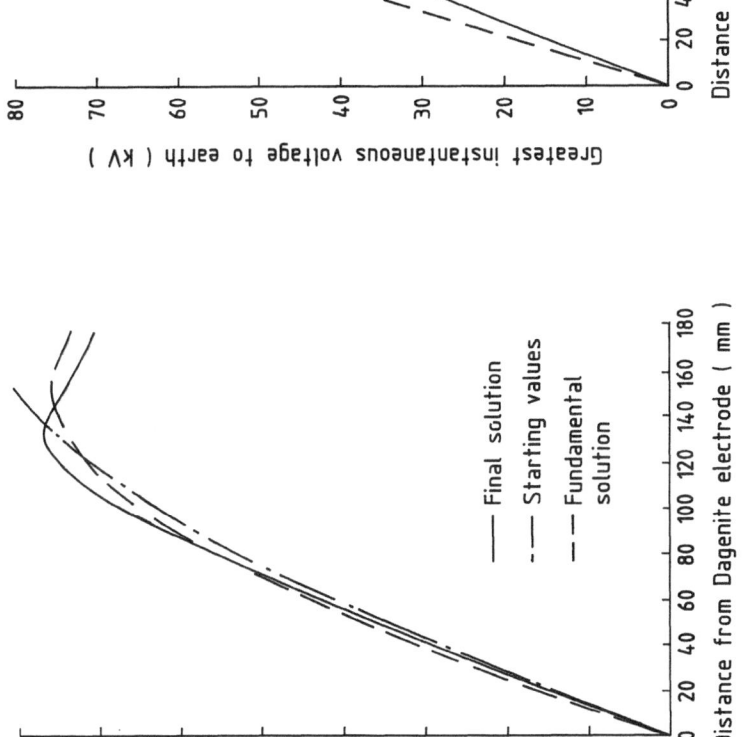

Figure 2. Comparison of three stages of
the computation at 50 kV rms.

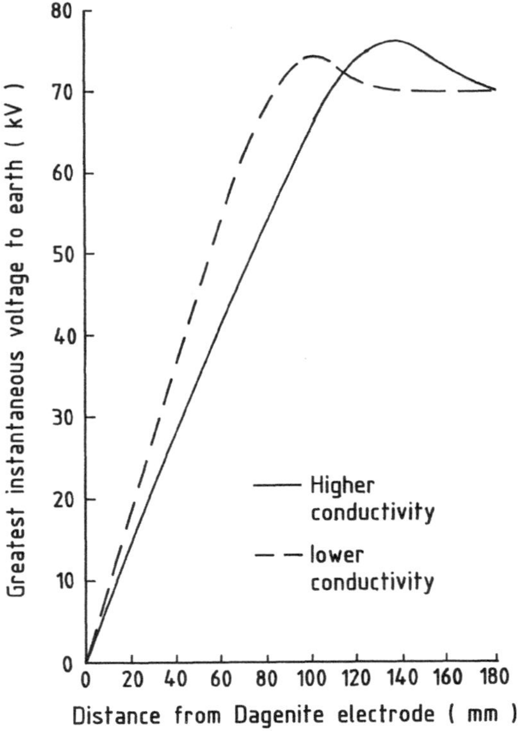

Figure 3. Effect of tape characteristic on
voltage distribution at 50 kV rms.

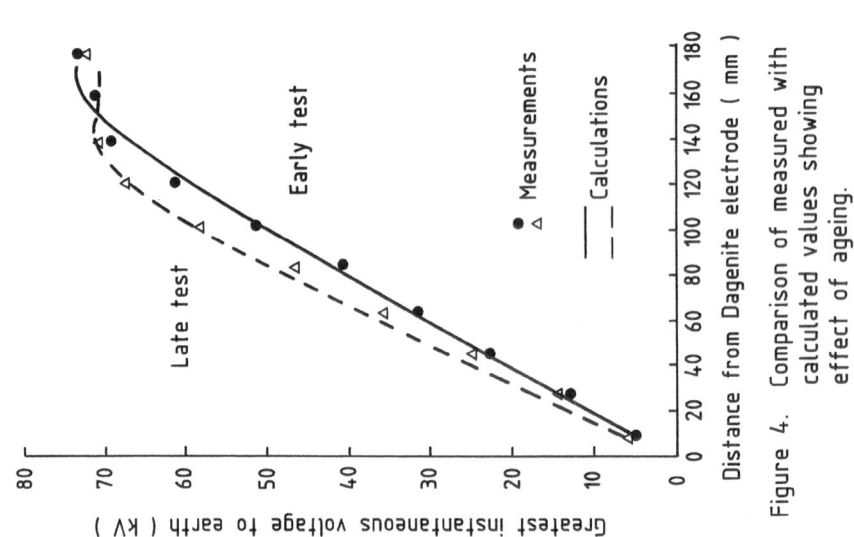

Figure 5. Comparisons with measurements for several voltages (early tests)

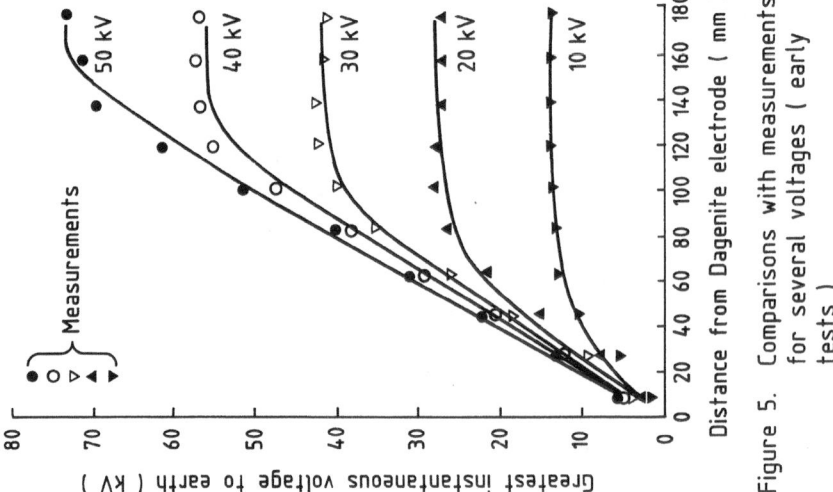

Figure 4. Comparison of measured with calculated values showing effect of ageing.

experimental results showed an ageing effect on the tape during tests. This raises a problem for theoretical predictions; how far might the tape have changed since its characteristics were measured?

Comparison with measurements. Figure 4 shows a comparison of calculated and measured voltages for a conductor bar voltage of 50 kV. The two sets of measured values were taken from the same tape before and after an extensive series of tests and show the effect of ageing. In the earlier test tape voltages were measured for a range of conductor bar voltages from 10 kV to 50 kV; comparisons with calculations are shown in Figure 5.

3. EDDY CURRENT PROBLEMS AND SURFACE IMPEDANCES

Physical background

Eddy currents at power frequencies in solid ferromagnetic bodies are usually confined to a narrow layer below the surface. Even for surface magnetising forces as high as 20000 A/m a depth of 5 mm is equivalent to infinity. It is possible to treat the eddy current phenomena in mild steel and other ferromagnetic materials as surface effects which link the normal component of flux density with the surface divergence of magnetising force (Preston and Reece, 1976). The term 'surface impedance' is used for the ratio of the complex fundamental of the surface electric force to the complex surface magnetising force; since the surface electric force is proportional to the time derivative of the total magnetic flux per unit area beneath the surface, the surface impedance contains all the information required to define a boundary condition for the magnetic fields. The use of surface impedances can lead to very rapid computer programs (e.g. Phemister et al. 1982) with no loss of accuracy in comparison with full numerical treatment of the eddy currents in the steel.

Application of the method to calculating surface impedances

Maxwell's equations, curl $\underline{H} = \underline{J}$ and curl $\underline{E} = -\frac{\partial B}{\partial t}$, with the displacement current neglected, reduce in one dimension to

$$\rho\frac{\partial^2 H}{\partial z^2} = \frac{\partial B}{\partial t} \text{ ,}$$

(5)

where ρ is the resistivity (taken here as constant),
z is the distance into the material,
H is the magnetising force, and
B is the flux density (a non-linear function of H).

When Equation (5) has been solved, the surface impedance can be computed as the ratio of the complex fundamentals of $\rho\frac{\partial H}{\partial z}$ and H at $z=0$.

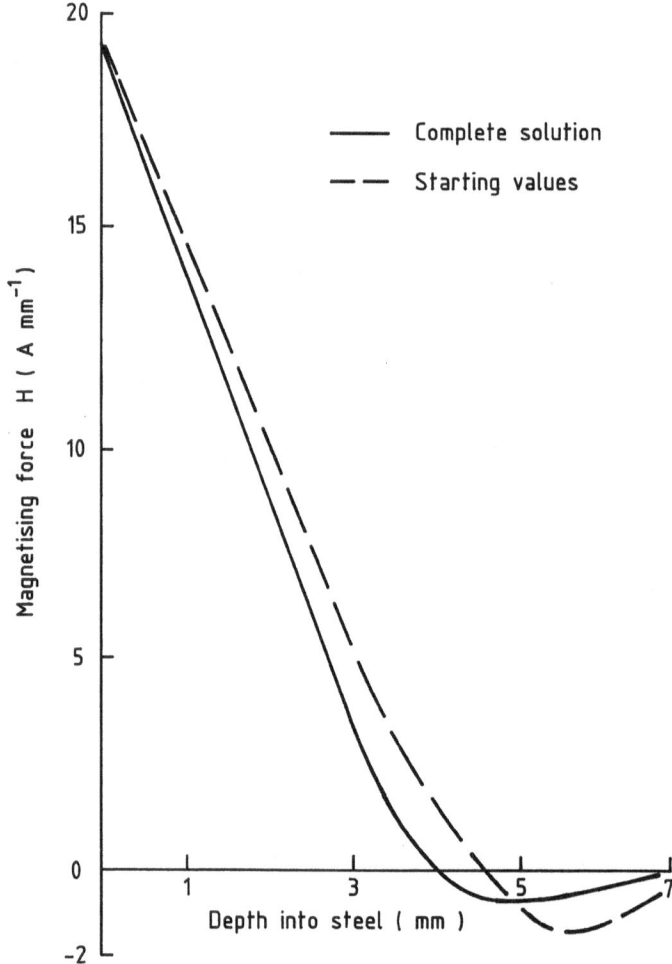

Figure 6. Calculations of magnetising force H at time of maximum at surface.

Equation (5) is similar in nature to Equation (1) and the methods used for stress-relieving tape seemed appropriate for a rapid computation of surface impedances. Stage one is irrelevant to this and, so far, only stage two has been fully programmed.

Results
Starting values of H found in stage two are compared in Figure 6 with final values from a very old and very slow program, which solves Equation (5) by forward iteration in time until all transients have largely decayed. It is clear that, when stage three has been completed, a much more efficient program will be available for computing surface impedances.

4. CONCLUSIONS

The four stage method can be summarised as follows:-

stage one uses a finite element technique to create boundary conditions imposed by the exterior linear regions,

stage two calculates a set of interior starting values in the fundamental harmonic by discretising the non-linear parabolic equation and assuming an exponential variation,

stage three is a straightforward iteration from the starting values to a full solution in the fundamental harmonic only,

stage four brings together the exterior and interior conditions with a hybrid of relaxation and Newton-Raphson to converge on a full harmonic solution.

The method could be applied to a wide range of non-linear problems and is a convenient way of combining linear and non-linear solutions. It brings together Fourier analysis, finite element procedures, finite differences, relaxation and Newton-Raphson iterative schemes. However the most important feature, which the author believes to be an innovation, is the high standard of starting values produced at the second stage. Any non-linear problem depends strongly on the quality of its starting values for efficient solution. In certain situations it may not be necessary to search for great accuracy, and the fundamental solution of stage three, or even in some cases the starting values themselves, may meet requirements. The method as a whole has proved to be a quick and reliable means of calculating a full harmonic solution to certain non-linear problems.

5. ACKNOWLEDGEMENTS

The work formed part of a joint programme of research by NEI Parsons Ltd. and the Central Electricity Generating Board. The author wishes to thank NEI Parsons for permission to publish, and Dr. D.A.H. Jacobs of the CEGB for his encouragement. The author is indebted to several colleagues at NEI Parsons, in particular Dr. J.W. Wood, Mr. B. Edwards Mr. R.T. Hindmarch, Mr. E.L. Slattery and Mr. T.G. Phemister.

6. REFERENCES

Phemister, T.G., Whitfield, A.H., and Jack, A.G., Field computation of subtransient damper impedances for use in optimal generator design, COMPUMAG Conference, Chicago, 1981, IEEE Trans. on Magnetics, Vol. MAG18, No. 2, March 1982, pp 633-637.

Preston, T.W, and Reece, A.B.J., The prediction of machine end-region fluxes, allowing for eddy current losses in thick components, Proceedings of 1st COMPUMAG Conference, Rutherford Laboratory, 1976, pp 213-220.

Wood, J.W., The properties of silicon carbide loaded resins and their application to electrical stress control in turbogenerators, Thesis for Ph.D., University of Strathclyde, 1982.

APPENDIX. EQUATIONS FOR STRESS-RELIEVING TAPE

The equation div $\left(\underline{J} + \frac{\partial \underline{D}}{\partial t} \right) = 0$, where $\underline{J}$ is the current density and $\underline{D}$ the electric displacement, follows directly from one of Maxwell's equations. If a two-dimensional problem is treated, integration of this equation across the thickness of the tape gives

$$\frac{\partial I}{\partial x} + \frac{\partial}{\partial t} \left(g \frac{\partial D}{\partial x} + D_{n1} + D_{n2} \right) = 0, \qquad (\alpha)$$

where I is the current per unit width along the tape,
g is the thickness of the tape,
D is the mean value of the component of $\underline{D}$ parallel to the tape, and
D_{n1} and D_{n2} are the outward components of electric displacement on either side of the tape. Although the behaviour of a semiconductor in an alternating field is more complicated, reasonable predictions can be made by taking $I = F(E)$, an odd function of electric force, and $D = \varepsilon_0 \varepsilon_1 E$, where ε_0 is the fundamental constant of permittivity and ε_1 is a constant relative permittivity.

The electric fields in and around the tape are three orders of magnitude higher than those induced electromagnetically in large generators and so the electric force can be treated as the negative gradient of a voltage, V, which will be taken as the voltage between the tape and the conductor bar. Since the insulation on the bar is thin (6 mm) compared with the scale of variation along the tape, it is sufficient to represent D_{n1} as $\varepsilon_0 \varepsilon_2 V/h$, where ε_2 is the relative permittivity of the insulation and h is its thickness. D_{n2}, the normal component of electric displacement on the outer surface of the tape can then be written simply D_n and Equation (α) includes

$$G \frac{\partial^3 V}{\partial x^2 \partial t} + H \frac{\partial}{\partial x} F\left(\frac{\partial V}{\partial x} \right) - \frac{\partial V}{\partial t} = H \frac{\partial D_n}{\partial t}, \qquad (\beta)$$

where $G = g h \varepsilon_1 / \varepsilon_2$ and $H = h/(\varepsilon_0 \varepsilon_2)$.

The function F is far from linear; it behaves exponentially.

SESSION D

SIMULATION & COMPUTING TECHNIQUES I

Chairman

A O MOSCARDINI

Sunderland Polytechnic

CAD ASPECTS OF ELECTROMAGNETIC FIELD CALCULATIONS

E.M.Freeman
ICST, London

1. INTRODUCTION

Many electrical machine designers would claim that there
is nothing new in CAE, since they have been using
computers for the last three decades for just this
purpose. Indeed, there are still those ancient stalwarts
who insist that "nothing is faster than their slide-rule".
However, times and techniques are changing, and it appears
to be generally agreed that with careful planning, modern
CAE methods can produce increased benefits and profits.

2. A BRIEF HISTORY

In 1882 Crompton laid down the foundations of classical
machine design theory. Later work by many others added
enormously to the body of literature on the subject.
However, the design process continued to be based on a
circuital model. Only discrete portions of the device were
subjected to any form of field analysis. Numerical methods
of field solving were almost unknown, although analogue
methods were extensively employed. By the late 1930's the
basis of the modern design method had been well established and described in the well known text books familiar
to all electrical machine designers. In the early days,
all calculations were performed by hand, slide rule,
mechanical calculator or graphically. The design process
was slow and time consuming. The advent of digital
computers in the late 1950's greatly increased the speed
of calculation, Veinott, 1972, lists over 100 references
to publications devoted to various aspects of CAD applied
to classical electromagnetic device design. Veinott's
book, early though it was, still provides much valuable
advice to would be computer aided designers.

3. THE PRESENT SITUATION

The development of computer based aids has been continuous
and steady. However, in parallel, there have been two
separate major lines of development, which will have an
important impact on the machine design process. These are
(i) high resolution interactive computer graphics, and
(ii) numerical methods for analysing magnetic and electric

field problems. in addition there has been the recent very heavy investment in CAE for electronics, and mechanical engineering design and manufacture. The mechanical stress analysis methods can be very advanced indeed. It is now common to find large CAE turnkey systems in everyday use for almost every aspect of engineering design, management control, production planning, manufacture and test. The application to the design of electrical machines is progressing somewhat more slowly. Fig. 1 shows a flow chart of a typical CAE process, with the classical electrical machine design portion indicated by Branch A.

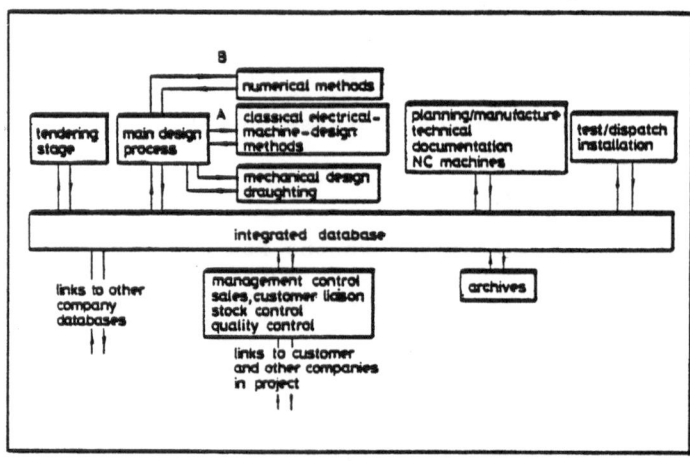

Fig. 1. The flow chart of a typical CAE process. Branch A represents the classical electrical machine design path. Branch B represents the parallel path in which numerical methods can be employed to analyse electromagnetic and thermal behaviour.

All the other portions are linked via the computer and share a common database. In the normal course of events, the electrical machine designer is responsible not only for the electrical design, but also for seeing that particular project through the whole tendering, design, production, manufacture, installation and on-site test process. Any problems usually end' up on his desk. However, with a few notable exceptions, Fig. 2, no computer link exists between the main computer CAE activity and the designer's electrical design programs. This means that the output from the designer's computer must be conveyed in some form to the main computer, for they are rarely the same, and it must be re-input with a

high chance of error. This takes time and militates
against any possibility of fast interaction between the
mechanical and electrical aspects of the design to obtain
an optimal solution. Furthermore, if the designer would
like to check his final version with one of the several
magnetic analysis packages available, eg, MAG-NET ELEVEN,
PE2D, or FLUX-2D, he must go to yet another computer,
usually to be found in the R & D section. This activity
is shown as Branch B in Fig.1, and in greater detail in
Fig. 2. We thus have a set of loosely connected
activities which could greatly benefit from being more
closely linked. Two points must always be borne in mind:
(i) no two people will want the same facilities; and (ii)
there is no substitute for the ingenuity of the good experienced electrical machines designer. We are simply
listing here a set of "tools" to improve his overall effectiveness in an increasingly competitive world.

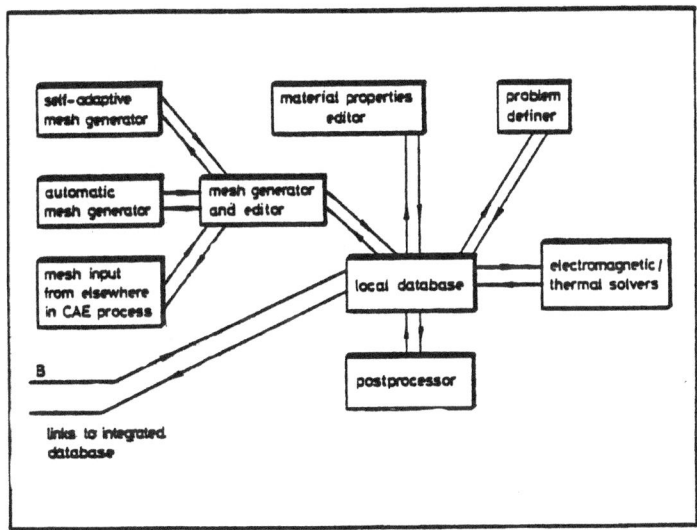

Fig. 2. Details of Branch B, of Fig. 1, showing a
possible scheme for the overall numerical analysis
process. Meshes may be input by hand, created automatically, or they can simply be taken from some other part of
the CAE process. The post-processor may be anything from a
simple plotter of curves, to a state-of-the-art package
capable of yielding any result the designer might require,
from flux densities to full performance predictions. Every
component shown links to the local database via standard
interfaces, thus giving maximum choice and flexibility.

The work station

Many companies have already invested in CAE. This can be anything from a small single work station, through to the larger installations to be found in the major electrical engineering companies. One UK electrical company has over 150 work stations, although only a handful are yet in use for the electrical engineering aspects of electrical machine design.

Most of the available CAE systems are sold as turnkey packages. The hardware is often specially designed for the purpose, particularly the high definition interactive colour graphics. It is mainly this feature which distinguishes the new CAE hardware, from the earlier computers. Hitherto, the designer held a model of the system in his mind. It was therefore limited in its physical complexity. Any alterations he might have made were to a fairly basic model, which might have been one of a large number from which he could select. With a modern CAE system he can be constantly presented with an updated display of the model. Briefly, a basic system will include: a 16/32 bit CPU, half a megabyte of memory; an alphanumeric terminal; a colour graphics display unit with at least 512*512 resolution; a 10/20 Mbyte hard disk plus a floppy disk drive; a digitiser tablet; a fast plotter for accurate draughting; a fast dot-matrix printer; a good operating system; and fast local area networking facilities.

There is no doubt that the introduction of fast local area network links will greatly add to the flexibility of using such systems. As an example, although there are many packages in use for the mechanical design of .electrical machines, there are, as yet, only a few which can be used to perform the detailed magnetic analysis. They nearly all require special dedicated hardware. The cost of converting the software can be very high. One easy solution to this problem is to simply have the two sets of different hardware connected by a fast local area network. The mechanical and electrical designers can then work side by side to arrive at the ever sought after optimal solution.

Use in the design office

It is becoming common to find quite advanced numerical methods in everyday use, usually by the R & D engineers, rather than the actual design engineers, although the latter are now taking a very close interest. The reasons include: increasing material costs; ever higher ratings; and ever more novel shapes. The classical design method, while well proven for conventional machines with reasonable ratings, cannot be expected to cope with such conditions. The only alternative is some form of numerical method. The popular one at present is based finite elements. Until recently the labour of inputting the mesh and examining the results has militated against its general use. Now, however, with fast pre-processors, the possibility of automatic meshing and input direct from other parts of the CAE process, using fast interactive graphics, these problems are largely obviated. Fig 2, shows an expanded form of Branch B of Fig. 1. The choice of method for mesh input to a mesh editor can be seen, this is followed by a problem definer, solver, and programmable post-processor, to produce the required results in a form familiar to the designer. The trend is towards standardised interfaces between the various components. This will make it possible for the designer to select and buy from a variety of software sources. It should not be overlooked that the writing of good, reliable, well documented software, in-house, can be very expensive.

With such facilities, the electrical machine designer can do his preliminary design using classical methods while cooperating with the mechanical designer on the mechanical and thermal aspects. Then, with minimal effort, he can perform a detailed magnetic analysis, and overall performance check. It should be noted that some designers are already adopting a blend of classical and numerical methods for all their magnetics design.

4. FACTORS TO BE CONSIDERED WHEN USING FE METHODS

The designer's task is to produce some form of electromagnetic device which will store and or convert energy. There will be a multitude of conflicting criteria to be satisfied. The design might be totally standard of fixed shape, requiring little effort. Alternatively, the problem might be totally open-ended, requiring a systems approach. In either instance there will be a point where it will be necessary to perform some form of electromagnetic analysis.

In the circuital approach, this is usually done by assuming a certain flux level; and then by working round a closed flux circuit, the total Ampere-turns may be determined. The number of magnetic circuit elements has traditionally been low. It was only necessary to postulate a flux in a pole, and allowing for the ever present leakage, progress round the circuit adding up the required ampere turns, to arrive at the total mmf. The leakage factors required faith and experience. The method works well until the shape is radically changed and/or the rating is pushed too high. Furthermore, it does not allow any form of experiment with the magnetic circuit shape, such as is required for optimisation, particularly weight reduction.

In the finite element model, the first most obvious factor is the large number of elements, commonly several hundred are used, the total could be several thousand. As a first approximation, there are twice as many nodes as elements, for first order elements; for a more precise estimation, see below. Each node has an associated potential. In a typical problem some of these will have specified values, but the majority will have to be calculated. The time required to solve a given problem increases as the number of unknowns to the power 1.5. Every effort should therefore be made to keep the number of unknowns to a minimum. Fortunately in many electromagnetic devices there exists a high degree of symmetry and/or repetition. These can both be exploited to reduce the initial model building and equation solving times. To assist this process it might be necessary to make certain pragmatically justified engineering assumptions. Hence the designer obtains a sub-model, which will accurately represent the whole. There should be no need to further sub-divide, as is the custom in the classical method, unless a special study of a portion of the model is required.

Rethinking and training required

It is common to find that the training designers need falls naturally into two phases, each requiring one or two days. in the first they learn of the finite element method, the information they need to have available before approaching the computer, and the methods by which they can extract results. In the second, hands-on, phase they learn how to use a specific set of software and hardware. Most software suppliers run such courses. Some of the factors involved in the first phase are described in the following sections.

The information required

In addition to the mesh itself, it is necessary to give other information in order for the problem to be fully specified. If certain information is not given, default values are usually assumed. On the other hand, the omission of key data could lead to a null result. The list of essential items will vary from system to system, depending on how well, or how badly the system has been designed. A complete list would include the following:

-Geometric information. The mesh; nodes with the coordinates, elements, material labels; boundary information, node constraints; the coordinate system used.

-Material properties. To include magnetisation curves, there might be several for a hard magnetic material; iron loss curves; conductivities, in time varying problems. These might all be drawn from a materials library, already input to the system.

-Problem definitions. A number of solutions might be required for problems which have the same geometric information, but differing material properties, excitations, or frequencies/time steps (if time dependent). The user might also want to control the degree of accuracy required in the solver, and the order in which the problems are solved, the order of the finite elements to be used. Finally he might want to look at intermediate results as the solver progresses, as a check on the process.

All this has to be input, or left set at the default values, before entering the solver phase. It is convenient to split the overall input process into three parts:
    1. A mesh editor
    2. A materials editor
    3. A problem editor
As an example, in the MAG-NET ELEVEN system, these are referred to as: MAGMESH, MAGCURV and MAGPROB, respectively.

The advantages of splitting the process in this manner are clear. The mesh might be used for a large number of materials, excitations, and under a variety of time varying conditions. The materials library need only be input and verified once ever, although it might be necessary to generate special material curves, which are problem dependent. The problem editor enables the user to specify a large number of problems with very little effort.

5. PRE-PLANNING

Much time and effort can be saved by pre-planning before going anywhere near the computer. The designer has to think about the form and best way of inputting the information about the problem. As a first step, the most important factor to decide is "what is the object of this exercise?". Is it a detailed field distribution, an

integral quantity such as impedance or torque, or is it
totally general problem? The choice can greatly affect th
whole approach to the modelling. Next a clear sketc
should be made of the physical device, including all thos
parts which might have some effect on the field distribu
tion. The sketch should show the coordinates, of eac
fiducial (key) point required for the final sub-model
together with a note about the units and coordinate syste
employed. N.B. many drawings only show the distance
between points. Future systems will no doubt allow fc
such information. Regions sharing common properties shoul
be clearly indicated. A list of the materials and thei
relevant properties is also required. Any material equiv
alents are also useful. The direction normal to the mai
plane of interest should be examined with a view to intrc
ducing approximations which can represent the model i
that normal direction. A note should be made of th
excitation current/flux levels and directions. Will ther
be a succession of of models differing only in that the
have slightly different material distributions?

Other factors which might have to be taken into accoun
include: the maximum number of nodes, for a given solver
the formulation method employed; the method for dealin
with open boundary problems; the possibility of sensibl
making use of another computer via a link, in order to d
the number crunching; the availability of a number o
solvers, each capable of solving a particular form o
problem; and finally, and so often neglected in some en
vironments, the financial cost of the operation.

At this stage the designer should be able to finalise th
following items:

-the approximations to be made and the smallest sub
problem, or set of sub-problems, to represent the whole;

-the boundary conditions for the actual model, or sub
model to be used;

-the form of mesh which will best yield the result
required, this might involve actually altering th
original physical model in the interests of economy an
expediency;

-the material properties and their dependence (if any) o
temperature, frequency, etc;

-the modelling method to be employed for non-linea
regions, eg permanent magnet regions;

-the excitation levels and time variation (if any);

-the type of solver and the accuracy required;

-the order of the finite elements to be used;

-the succession of problems to be solved in the order required.

All this is to be done before going approaching the computer. We will next examine some of the above listed points in detail.

The reduction of the model to a sub-model

In the normal classical design process, the designer makes many implicit assumptions during the course of his analysis. In the finite element process it is necessary to be absolutely clear about any assumptions, they all have to be declared somewhere in the input process; either by the user, or indirectly as default options set by the system designer. The first of these assumptions occurs when reducing the model to the minimum sub-model. If we take as an example a salient pole machine, it is obvious that the pole structure repeats itself every pole pitch. Similarly, the tooth/slot pattern on an induction motor, or synchronous machine, stator repeats every slot pitch. The excitation, on the other hand, repeats every pole-pair for the salient pole machine, and may not repeat at all for the ac machine. The important point here is that there are generally two forms of pattern to be noted. The first is for the structure, and the second for the excitations. When assembling a sub-model it is necessary to consider both.

The sub-model and the boundary conditions

A good example in a machines context is the salient pole machine, dc or ac. The pole structure pattern repeats every pole pitch. Furthermore, the current patterns will usually repeat in the same way. Therefore under normal conditions it is necessary to model only one pole pitch. The test is simply this, "does the chosen portion look similar, allowing for alternate current directions, to all the equivalent portions elsewhere in the device?". If so, then the chosen portion is a valid one.

Similar arguments apply to isolated models set in infinite space and models which can be regarded as one of a a repeated set. For an isolated model the user must specify some form of limiting boundary somewhere. At the simplest level the user has the choice of two possibilities. He can either assume that the original model is placed within box made of (a) iron of infinite permeability; or (b) su perconducting material. If the former, then all flux line will meet the boundary 'normally'. If the latter, then n flux line will cross a boundary. If one of a repeated set then (a) and/or (b) might both apply.

Many cylindrical geometry problems can be simplified i the field variation in the azimuthal direction is zero, o cisoidal, ie the problem is axisymmetric. The user mus specify:

- that the problem is axisymmetric
- the form of the azimuthal variation
- and also the location of the z-axis.

Thus, so far there are three types of boundary condition needed to specify what happens on the boundaries o typical field problems. Two others commonly encountere are: infinite boundary; and surface impedance.

The mesh when constructed will have elements and nodes. I will be necessary to label some of these nodes in order t specify the various boundary conditions. Summarising thes are:

-A constant magnetic vector potential (MVP) boundary requiring unary constraints, this is a Dirichlet boundary

-A boundary where the flux lines are normal, and therefor there is no tangential flux. This is a Neumann boundary.

-A pair of boundaries where the pairs of MVP values ar linked in some simple way, eg pole centre-lines; binar constraints are used here.

B-H curve and permeability modification
The designer is continually trying to simplify problem while retaining all the important key features. He i trying to make make them tractable, and fast and inexpen sive to solve. Often he must employ techniques, which ca only be pragmatically justified. Examples occur i machines analysis as a result of modelling the three dimensional problem in 2-D.

There are several forms of anisotropy which can occur i magnetic field problems; some due to the materials, an others due to the structure. Material anisotropy has tw forms, for soft and hard magnetic materials respectively. For soft materials, some steel manufacturers supply mag

netisation and loss curves for the rolling direction and
normally to the rolling direction. Not only must the
solver accommodate the anisotropy, but also there should
be a simple method in the pre-processor for indicating the
preferred, or rolling direction. This problem is gaining
increasing importance as more designers choose to use
grain-oriented steels. For a further discussion of this
problem the reader is referred to the book by Ferrari and
Silvester.

Equally important is the problem of accurately modelling
permanent magnet materials. There is very little hard data
available in this area. Some designers ignore all effects
normal to the main direction of magnetisation; others
allow some effect; a limited few actually attempt to
gather experimental data for their modelling. The work by
Binns is to be noted in this context.

Excitation in polyphase machines

Modelling can often be simplified if the winding is
replaced by the fundamental current distribution, ignoring
thereby all mmf harmonics. Obviously this approximation
may only be used if the main area of interest is away from
the tooth/slot regions. Where this is difficult to
achieve, within a pole-pitch, because of a non-integral
slot/pole value, it is occasionally possible to replace
the actual slotting with an equivalent slotting. This is
done by keeping the iron/air ratio unaltered to maintain
the tooth density at a realistic value. This is permis-
sible only if the subsequent change in slot/gap ratio is
negligible.

6. CONCLUSIONS

The use of standard data interfaces will greatly increase
the ease of introducing new software packages. Among
these, are: permanent magnet solvers; analyses for special
forms of rotary machine; optimising methods; and coupled
system analyses.

All the above comments have been made in the context of 2-
D problems. The everyday use of full 3-D CAD methods in
the design office is still rare.

The market for the designer's products is large, but very
competitive. In the future the designer will need to make
use of every possible available aid in order to prosper.
The new methods will have to be well tested and proven, to
satisfy all quality assurance criteria. The methods will

not be the panacea for all ills, but they will help. Adequate training will be essential.

The introduction of CAE will alter the structure and organisation of industry. However, the most important single factor will always be the combination of the creativity, ingenuity and experience of the designer.

The biggest change will be in the way people think.

7. BIBLIOGRAPHY

Csendes, Z.J., Freeman, E.M., Lowther, D.A. and Silvester, P.P., "Interactive computer graphics in magnetic field analysis and electric machine design", IEEE Winter Power Meeting (Feb.1981)

Freeman, E.M. and Trowbridge, C.W., "Interactive computer graphics for CAD", Electronics and Power, April 1982, pp331-334.

Freeman, E.M., "Interactive computer aided design of electric machines and electromagnetic apparatus", Invited paper for the Intermag Montreal, July 1982, JAP, Nov. 1982, pp8393-8398.

Raby, K.F., "On getting the right answer", Chairman's Address, SET Division, Proc.IEE. (Jan. 1982).

Silvester, P.P. and Ferrari, R.L., "Finite elements for electrical engineers", CUP, 1983.

Silvester, P.P., Lowther, D.A. and Freeman, E.M., "Finite element mesh generation using a small computer with interactive raster graphics", ICEM, Athens (Sept.1980).

Tarkanyi, M., Freeman, E.M., Silvester, P.P. and Lowther, D.A., "The interface between finite element methods and machine design", IEE EDAM Conference, July 1982.

Veinott,C.G.,"CAD of Electric Machinery",MIT Press, Cambridge Mass., 1972.

"Computer Aided Design of Electromagnetic Devices", Short Course, Electrical Engineering Department, Imperial College, January, 1983.

The interested reader is also referred to the proceedings of: Compumag 1976, 1978, 1981, and the annual Intermag Conference.

SIMULATION OF HYSTERESIS IN FERROMAGNETIC MATERIALS

D O'Kelly

School of Electrical Engineering, University of Bradford

1   INTRODUCTION

In many electromagnetic field problems the hysteresis effect present in the ferromagnetic materials may be neglected without significant loss of accuracy. However, hysteresis must be included in the modelling of permanent-magnetic-type materials or with any ferromagnetic material when the hysteresis loss is a substantial part of the total core loss. A wide range of problems require the dynamic modelling of hysteresis. These cover the whole spectrum of material properties from residual-flux and saturation effects in current transformers using low-loss steels to recoil characteristics in hard steel permanent-magnet field systems.

The inclusion of hysteresis in magnetisation characteristics has been represented by several different approaches. Trapezoidal loop approximations (Copeland et al, 1963), analytic functions (Davis, 1971), the use of multi-point characteristics of the measured B/H loop family (Zakrzewski et al, 1971) may be used. All these methods are restricted to flux excursions due to steady-state cyclic magnetisation.

The simulation of transient performance requires the mathematical model to evaluate flux excursions for magnetising force changes not related to steady-state cyclic energisation including such features as recoil characteristics, residual flux effects, offset magnetisation etc. A limited amount of work has been done in this area (Wright et al, 1974; Talukdar et al, 1976; Janssens, 1976; MacFadyen et al, 1974; Coulson et al, 1977).

Two methods of modelling hysteresis have been used by the writer:
A. Nonlinear complex permeability. This technique is limited to problems with sinusoidal magnetisation and steady-state solutions.
B. A computer simulation which is suitable for problems with non-sinusoidal or transient magnetisation.

The paper gives a description of each method and briefly discusses their features and illustrates the range of application with typical examples.

2   HYSTERESIS MODEL A - COMPLEX PERMEABILITY

If a sinusoidal magnetising force $\tilde{H}_f \sin \omega t$ is applied to a typical B/H

115

hysteresis loop, Figure 1, the resulting flux density waveform is non-sinusoidal. The fundamental component of flux density $\overline{B}_f \sin(\omega t - \delta_h)$ may be related to the magnetising force (with both expressed in phasor terms) by a complex coefficient $\overline{\mu}$ (Butler et al, 1948; Aspden, 1952).

$$\overline{B}_f = \overline{\mu} \, \overline{H}_f \tag{1}$$

where

$$\overline{\mu} = \mu \exp(-j\delta_h) = \text{complex permeability}$$

$$\delta_h = \text{hysteresis angle.}$$

The hysteresis loop defined by the fundamental components of B_f and H_f is elliptical with an area equal to that of the original B/H loop. Hence, the hysteresis loss is unchanged. Applied to a family of B/H loops, nonlinear characteristics of μ and δ_h as a function of H_f can be derived. Figure 2 shows typical characteristics for a medium-hard material (mild steel).

Application
Since all harmonic effects are neglected this method of representation may relate sinusoidal changes of magnetisation and flux density in ferromagnetic material in both time and space. The application is limited to simple geometries in which the material may be conveniently sub-divided into planar or circular layers (O'Kelly, 1975a). Each layer is assumed to have a value of $\overline{\mu}$ dependent upon the average peak magnetising force in the layer and is represented by an equivalent electrical analogue in the form of a T or π circuit.

Figure 3 shows the equivalent-T network for one layer of a semi-infinite lamination with (a) alternating magnetisation and (b) travelling wave magnetisation.

A feature of this model is that with travelling wave excitation different values of complex permeability in the two orthogonal directions can be included (Freeman, 1967). Furthermore, the different loss characteristics associated with rotational and pulsating fields (Brailsford, 1968) can be simply represented (O'Kelly, 1976b).

3   HYSTERESIS MODEL B - COMPUTER SIMULATION

Many different models were examined before concluding that the following basic technique contained the essential features for the modelling of both steady-state and transient B/H excursions (O'Kelly, 1977c). Figure 4 illustrates the method. The family of B/H loops is defined by three parameters:
(a)   a nonlinear curve
(b)   a width factor W
(c)   an exponential factor which defines the excursion between two points.

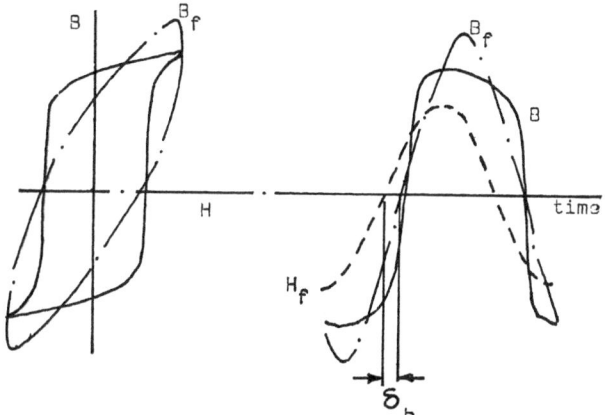

Fig. 1 Complex permeability

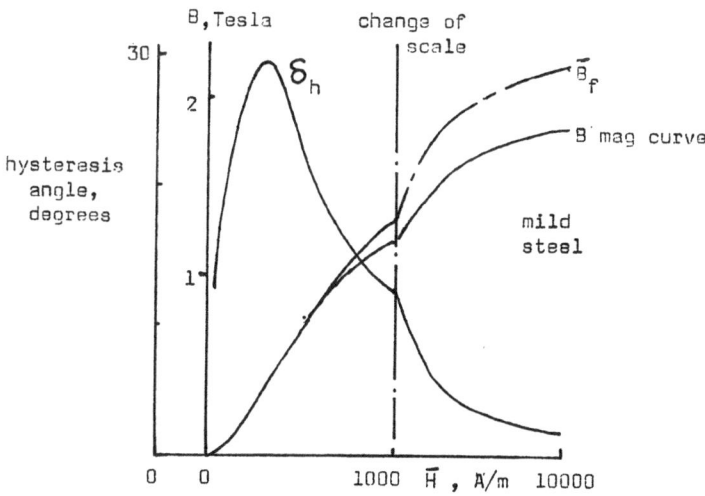

Fig. 2 Non-linear complex permeability characteristics

117

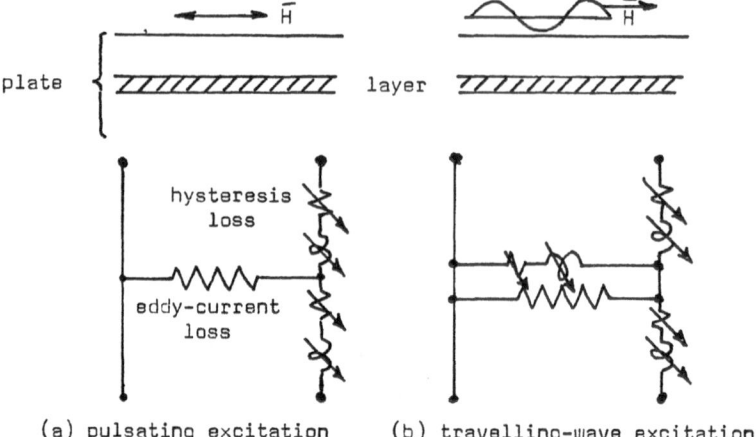

(a) pulsating excitation   (b) travelling-wave excitation

Fig. 3 Concentrated-parameter equivalent-circuit analogue
of layer of ferromagnetic material

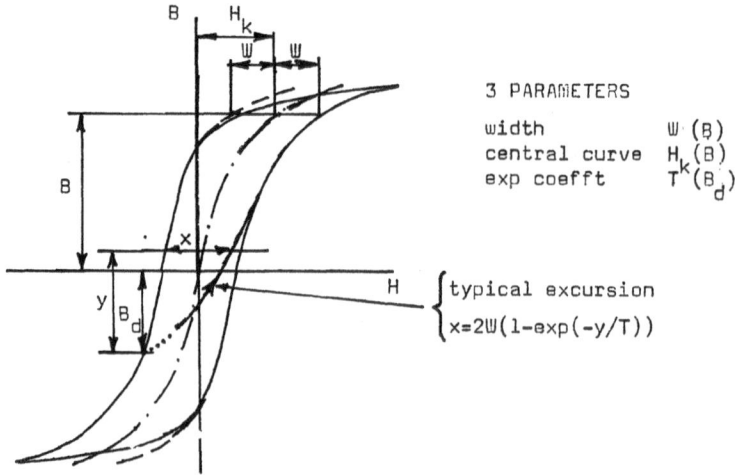

3 PARAMETERS

width $W(B)$
central curve $H_k(B)$
exp coefft $T(B_d)$

typical excursion

$x = 2W(1 - \exp(-y/T))$

Fig. 4 Computer simulation of B/H hysteresis loop

This method of representation is generally applicable to a wide range of hysteretic materials. Good agreement was obtained between the measured family of steady-state B/H curves and those computed for low-loss, medium-hard and hard materials (O'Kelly, 1977c). Although the accuracy for transient excursions is lower the technique permits residual flux, recoil characteristics, non-sinusoidal and transient magnetisation to be modelled.

The computation of flux waveshapes, losses etc in a lamination with alternating magnetisation uses a similar model to that for the complex permeability representation. However, for travelling-wave excitation a three-dimensional concentrated-parameter model is necessary, and the computing times are large.

4   ALTERNATING MAGNETISATION OF MILD STEEL PLATE

Figure 5 shows normalised results for the steady state loss in a mild steel plate energised with a sinusoidally-varying magnetising force (O'Kelly, 1980d). Mild steel was chosen because a wide range of specimen sizes from the same batch of material was possible and the hysteresi loss was measured and found to be independent of frequency up to 400Hz. The peak magnetising forces of 973 and 5838 A/m correspond to non-saturated and saturated values. The classical loss characteristic assumes the flux density remains constant across the plate and is unchanged by eddy-current action. Representative calculated results for a non-sinusoidal magnetising force (Figure 6) show the influence of the phase relationship of the harmonic and fundamental on lamination loss. Figure 7 shows flux penetration with impact excitation.

5   HYSTERESIS MOTOR SIMULATION

Measured and computed results for a hysteresis motor are shown in Figure 8. These representative results (O'Kelly, 1976e) show that with the complex permeability modelling technique the influence of rotational hysteresis can be included to give improved accuracy of simulation.

6.   CONCLUSIONS

Two methods are described to model hysteresis in ferromagnetic material represented by an equivalent (concentrated-parameter) electrical analogue. The complex permeability technique is relatively easy to apply and the effects of rotational hysteresis, different permeabilities in orthogonal directions etc can be simply included. The second method using a computer model is more accurate, providing information on losses, flux density waveshapes etc in a lamination with pulsating excitation at the expense of higher computing times. The application to three-dimensional fields is under investigation directed towards reducing the computation time and extending the range of application.

119

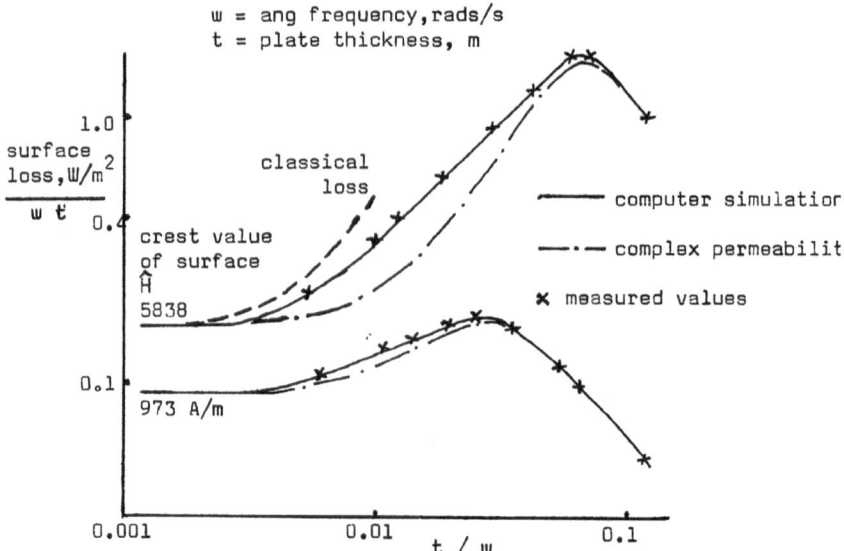

Fig. 5   Normalised loss in mild steel plate

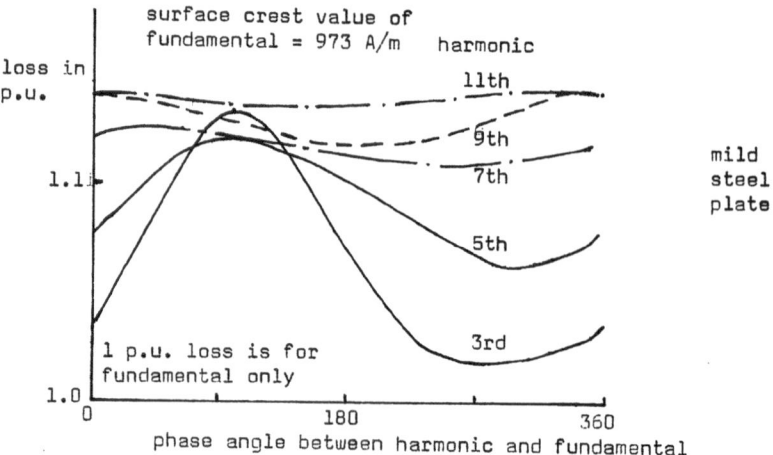

Fig 6   Influence of phase angle on loss of fundamental and
20 % harmonic content

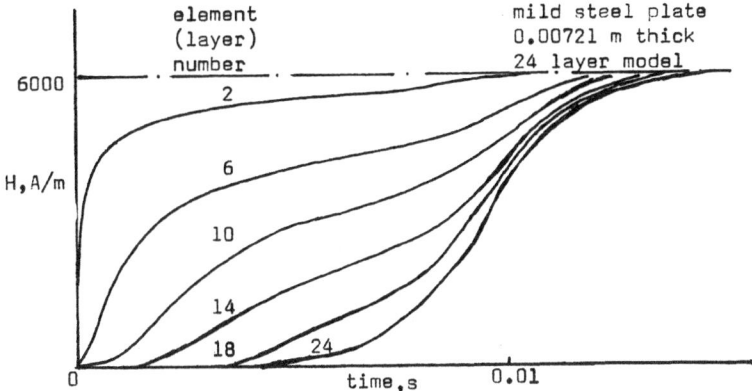

Fig 7 Penetration into mild steel plate with impact
excitation

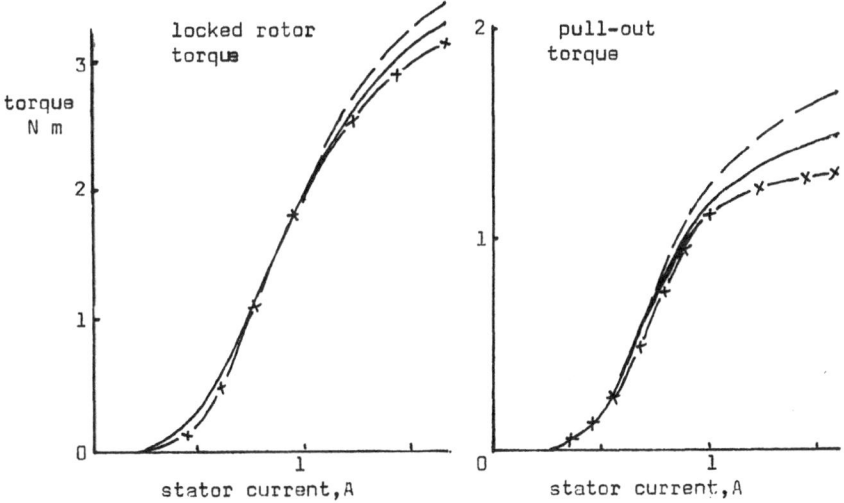

Fig 8 Hysteresis motor torque characteristics

    ——————— computed (rotational hysteresis effect included)
    — — normal computed values
    —x— measured values

121

REFERENCES

Aspden, H H "Eddy current in solid cylindrical cores using non-uniform permeability", Jrnl.of App. Physics, 25, pp. 523-528, 1952.

Brailsford, F "Rotational hysteresis loss in electrical sheet steels", Jrnl. IEE, 83, pp. 566-572, 1938.

Butler, O L & Mang, C Y "Predetermination of the magnetic properties of ferromagnetic lamina", Jrnl.IEE, V95, Pt.II, pp. 25-35, 1948.

Copeland, M A & Slemon, G R "An analysis of the hysteresis motor", Trans. IEEE, PAS-82, pp. 34-41, 1963.

Coulson, M A; Slater, R D & Simpson, R R S, "Representation of magnetic characteristics including hysteresis using Preisach's Theory", Proc.IEE, 124, pp. 895-898, 1977.

Davis, N, "Derivation of an equation to the B/H loop", J.Phys.D., pp. 1034-1039, 1971.

Freeman, E M, "Travelling waves in induction machines; input impedance and equivalent circuit", Proc.IEE, 114, pp.1681-1683, 1967.

Janssens, N, "Mathematical modelling of magnetic hysteresis", Proc. Compumag, pp. 191-197, 1976.

MacFadyen, W K; Simpson, R R S; Slater, R D & Wood, W S, "Representation of magnetisation curves including hysteresis by exponential series', Proc.IEE, 121, pp. 1019-1020, 1974.

O'Kelly, D, "Losses in cylindrical cores including hysteresis and eddy-current effects", J.Phys.D., 8, pp. 568-575, 1975.

O'Kelly, D, "Flux penetration and core loss in solid iron", IEEE Trans MAG-11, pp. 55-60, 1975.

O'Kelly, D, "Simulation of transient and steady-state magnetisation characteristics with hysteresis', Proc. IEE, 124, pp. 578-82, 1977.

O'Kelly, D, "Flux penetration and losses in steel plate with sinusoidal magnetisation", Proc.IEE, 127, Pt.B, pp. 287-292, 1980.

O'Kelly, D, "Theory and performance of solid-rotor induction and hysteresis machines", Proc.IEE, 123, pp. 421-428, 1976.

Talukdar, S N & Batley, J R, "Hysteresis models for system studies", Trans.IEEE, PAS-95, pp. 1429-34, 1976.

Wright, A & Carneiro, S, "Analysis of circuits containing components with cores of ferromagnetic material", Proc. IEE, 121, pp.1579-81,1974.

Zakrzewski, K & Pietras, F, "Method of calculating the electromagnetic field and power losses in ferromagnetic materials taking into account magnetic hysteresis", Proc. IEE, 118, pp. 1679-1685, 1971.

PREDICTION OF STABILITY MODELS USING

FREQUENCY RESPONSE DATA

A H Whitfield

Department of Engineering Mathematics
Loughborough University of Technology

1. INTRODUCTION

The system transfer function provides the engineer with a useful
method of predicting the time response and stability of a linear
system. For certain phenomena it may be relatively easy to model the
system by one or more differential equations and deduce the transfer
function by Laplace transformation. However, due to the inherent
complexity of some systems e.g. distributed parameter, or because of
the lack of precise knowledge of parameter values, the *a priori*
modelling of the system in the time domain may be impractical. In
such circumstances the transfer function must be predicted from
specified test data.

It is known that when a sinusoidal input of the form $A \sin \omega_k t$ is
applied to a system with transfer function $G(s)$ then the output from
the system is of the form $B_k \sin(\omega_k t + \phi_k)$ where

$$B_k = A|G(j\omega_k)|$$

$$\phi_k = \tan^{-1}\{\text{Im}[G(j\omega_k)]/\text{Re}[G(j\omega_k)]\}$$

The complex function $G(j\omega)$ is known as the frequency response
function. A set of frequency response data consists of complex
values $\{G(j\omega_k); k = 1, \ldots, M\}$ deduced from the output amplitude B_k
and phase shift ϕ_k observed at given input frequencies
$\{\omega_k; k = 1, \ldots, M\}$. Various external factors and measuring
inaccuracies will usually mean that the measured amplitudes and
phases will not concur precisely with the true values expected from
the system transfer function $G(s)$. It is the purpose of this paper
to outline various methods of finding the 'best' approximation to
$G(s)$ from the measured data.

In the paper three fundamental expressions for an approximating
system transfer function will be considered

Form I:
$$H(s) = \frac{a_o + a_1 s + \ldots + a_m s^m}{1 + b_1 s + \ldots + b_n s^n}$$

(1)

$$\text{Form II:} \quad H(s) = K \frac{\prod\limits_{i=1}^{m_1} (1 + T_i s) \prod\limits_{i=1}^{m_2} (1 + P_{1i}s + P_{2i}s^2)}{\prod\limits_{\ell=1}^{n_1} (1 + S_\ell s) \prod\limits_{\ell=1}^{n_2} (1 + Q_{1\ell}s + Q_{2\ell}s^2)} \qquad (2)$$

$$\text{Form III:} \quad H(s) = K \frac{\prod\limits_{i=1}^{m} (1 + T_i s)}{\prod\limits_{\ell=1}^{n} (1 + S_\ell s)} \qquad (3)$$

The parameter sets $(a_o, a_1, \ldots, a_m; b_1, \ldots, b_n)$,
$(K; T_1, \ldots, T_{m_1}; P_{11}, \ldots, P_{1m_2}, P_{21}, \ldots, P_{2m_2}; S_1, \ldots, S_{n_1};$
$Q_{11}, \ldots, Q_{1n_2}, Q_{21}, \ldots, Q_{2n_2})$, $(K; T_1, \ldots, T_m; S_1, \ldots, S_n)$
each being real.

Forms I and II are the most general and are equivalent. The principle methods for finding the parameter set of form I will be outlined in section 2 and implementation via a recursive least-squares approach will be demonstrated.

Form III is of particular interest in the determination of various parameters in synchronous generators. Given certain frequency response data it is desirable to characterise such properties as direct- and quadrature-axis impedance and inductance by equivalent circuits with positive time constants. Hence it is required to find an approximating transfer function of form III with $T_i > 0$ $(i=1, \ldots, m)$ and $S_\ell > 0$ $(\ell=1, \ldots, n)$. Section 4 will outline some approaches to this latter problem.

2. APPROACHES TO THE GENERAL FORM PROBLEM

The response of a system having a transfer function of form I to a sinusoidal input with frequency ω is characterised by

$$H(j\omega) = \frac{N(j\omega)}{D(j\omega)} = \frac{a_o + a_1(j\omega) + \ldots + a_m(j\omega)^m}{1 + b_1(j\omega) + \ldots + b_n(j\omega)^n}$$

126

A general error criterion representing the difference between the model predicted frequency response and the measured frequency response at the discrete set of frequencies $\{\omega_k; \; k = 1, \ldots, M\}$ is defined as

$$
E = \sum_{k=1}^{M} w_{k1}^2 \left\{ Re \left[\frac{N(j\omega_k)}{D(j\omega_k)} - G(j\omega_k) \right] \right\}^2
$$

$$
+ w_{k2}^2 \left\{ Im \left[\frac{N(j\omega_k)}{D(j\omega_k)} - G(j\omega_k) \right] \right\}^2 \tag{4}
$$

where w_{k1}, w_{k2} are specified weights. It is required to minimise E with respect to $(a_0, a_1, \ldots, a_m, b_1, \ldots, b_n)$ i.e. $(\underline{a}^T, \underline{b}^T)$. The presence of the unknown parameters $\underline{b}^T$ in $D(j\omega_k)$ makes the above a non-linear least-squares problem which is best tackled by a numerical optimisation technique. Various methods to avoid such a numerical approach have been proposed.

Method of Levy (1959)
This method is based on minimising the error criterion

$$
E' = \sum_{k=1}^{M} |N(j\omega_k) - G(j\omega_k)D(j\omega_k)|^2
$$

with respect to $(\underline{a}^T, \underline{b}^T)$. Thus denoting

$$
e_k = N(j\omega_k) - G(j\omega_k)D(j\omega_k)
$$

and noting that e_k is a complex number which is linear in the parameters of $\underline{a}^T$ and $\underline{b}^T$, then Levy's method is reposed as

$$
\min_{\underline{a}^T, \, \underline{b}^T} \sum_{k=1}^{M} Re[e_k]^2 + Im[e_k]^2
$$

This problem is now of the linear least-squares type and the parameters $(\underline{a}^T, \underline{b}^T)$ are given by solving the appropriate normal equations, provided no ill-conditioning exists. Referring to the criterion E, this latter criterion has $w_{k1} = w_{k2} = |D(j\omega_k)|$ and if $D(j\omega)$ varies significantly with ω then the values of $(\underline{a}^T, \underline{b}^T)$, computed from this weighting of the criterion E, will be unsatisfactory in general.

127

Method of Sanathanan and Koerner (1963)

An alternative error criterion was proposed by Sanathanan and Koerner to compensate for the weighting introduced by Levy's method while still retaining the linear least-squares structure. At an 'outer loop' iteration r ($r = 1, 2, \ldots, r_{max}$) an error criterion E_r'' is defined as

$$E_r'' = \sum_{k=1}^{M} \frac{1}{|D_{r-1}(j\omega_k)|^2} \; |N(j\omega_k) - G(j\omega_k)D(j\omega_k)|^2 \qquad (5)$$

i.e.

$$E_r'' = \sum_{k=1}^{M} \frac{1}{|D_{r-1}(j\omega_k)|^2} \; \{Re[e_k]^2 + Im[e_k]^2\}$$

where $D_{r-1}(j\omega_k)$ is the complex number obtained by evaluating $D(j\omega_k)$ using the values of $b_1, \ldots, b_n$ from the previous iteration. Either a specified number of iterations are performed i.e. r_{max} is fixed, or iteration continues until $(\underline{a}^T, \underline{b}^T)$ is judged to have converged. At any iteration r the linear least-squares problem is solved from the appropriate normal equations, once again assuming no ill-conditioning exists.

Method of Lawrence and Rogers (1979)

Using equation (5) as a basic error criterion Lawrence and Rogers have reformulated the problem as posed by Sanathanan and Koerner in a recursive scheme, avoiding the need to explicitly invert the normal equations. The equations which are necessary to implement this latter scheme are somewhat involved and are very problem specific. An equivalent recursive implementation which relies on more generally available, or easily written, software will now be derived.

3. RECURSIVE LEAST-SQUARES IMPLEMENTATION

As described in Appendix A, and in greater detail by Young (1974), the recursive least-squares (RLS) algorithm solves, without recourse to matrix inversion, the problem

$$\min_{x_1, \ldots, x_q} \sum_{i=1}^{p} \left(\sum_{j=1}^{q} c_{ij}x_j - d_i \right)^2$$

where c_{ij} ($i = 1, \ldots, p$; $j = 1, \ldots, q$) and d_i ($i = 1, \ldots, p$) ($p > q$) are supplied 'data' values. This algorithm is popular in discrete time systems identification and large identification packages are likely to contain a RLS module. A general error criterion and problem at an outer loop iteration r ($r = 1, \ldots, r_{max}$) to which the RLS algorithm is applicable are now given as

$$\min_{(\underline{a}^T, \underline{b}^T)} \quad E_r^* = \sum_{k=1}^{M} w_{rk1}^2 \operatorname{Re}[e_k]^2 + w_{rk2}^2 \operatorname{Im}[e_k]^2 \qquad (6)$$

By inspection since e_k is linear in $(\underline{a}^T, \underline{b}^T)$ it is seen that this problem is of the correct form for RLS solution, the precise formulation being detailed in Appendix B. Thus the previously outlined methods are seen to be contained in this formulation.

Levy : $r_{max} = 1$, $\qquad w_{rk1} = w_{rk2} = 1$

Sanathanan and Koerner : $w_{rk1} = w_{rk2} = \dfrac{1}{|D_{r-1}(j\omega_k)|}$

The recursive solution of the Sanathanan and Koerner method i.e. the proposition of Lawrence and Rogers is also included, though the algebra showing the equivalence is too tedious to be demonstrated here.

Problems encountered

As noted elsewhere (Payne, 1970) and as experienced by the author the method of Sanathanan and Koerner, and hence any recursive implementation of this method, may, for noise corrupted data, not yield even realistic results e.g. negative time constants for stable systems. Additionally, the nature of convergence of $(\underline{a}^T, \underline{b}^T)$ after successive outer loop iterations must be considered. Specifically, when the difference between the model numerator and denominator polynomial degrees i.e. $(m - n)$ does not agree with the corresponding difference of system generating the frequency response data, then the highest powers of s in $H(s)$ causing the discrepancy are adjusted by the algorithm to small though usually non-zero values. Such small values are not stable from one outer loop iteration to the next and hence convergence by all the parameters in $(\underline{a}^T, \underline{b}^T)$ may not be achieved.

Modifications

A new choice of error criterion within that of the general linear least-squares criterion (6) was found to give realistic results more frequently than the criterion of Sanathanan and Koerner, and Lawrence and Rogers. This new criterion is given by the selection of weights as

$$w_{rk1} = \frac{1}{\operatorname{Re}[G(j\omega_k)]\,|D_{r-1}(j\omega_k)|}$$

$$w_{rk2} = \frac{1}{\operatorname{Im}[G(j\omega_k)]\,|D_{r-1}(j\omega_k)|}$$

Clearly this latter criterion is based on the sum of relative errors squared of the real and imaginary components.

The convergence problem was overcome by finding the roots of both the numerator and denominator polynomials and therefore the system time constants. The discrepancy in the difference of polynomial degrees was found to give rise to 'negligible' time constants i.e. time constants which have a negligible effect on the model transfer function in the range of testing frequencies used. If one or more such time constants were noted then the numerator or denominator polynomial degree was appropriately reduced and the estimation procedure was repeated using the 'significant' time constants found previously to give corresponding initial values for $(\underline{a}^T, \underline{b}^T)$.

Consideration must be given to two other facets of the model numerator and polynomial degrees. In general the degrees of these polynomials will not concur with the corresponding degrees of the actual system transfer function. While the difference in degrees was accommodated above there are still the problems of choosing the degree a) too high b) too low. A discussion of these problems is given in the next section where it is also applicable.

4. MODELS REQUIRING REAL POSITIVE TIME CONSTANTS

In certain circumstances it is known that a suitable transfer function model is provided by that of form III, equation (3), wherein it is further desired that $K > 0$, $T_1, \ldots, T_m > 0$ and $S_1, \ldots, S_n > 0$. An example of this situation is in the area of synchronous machine parameter estimation where it is desired to find such models for several functions e.g the impedance function $Z(s)$, the inductance function $L(s)$ and the rotor/stator current transfer function $G(s)$.

If the supplied data were free from noise and measurement error then the expectancy is, and experience has shown, that the methods of the last section, provide accurate parameter values for correctly assumed model structures. However, in practice the experimental data may well be corrupted and while the new method introduced at the end of the last section has proved to be quite robust, it is desirable to use a method which is entirely robust and guarantees a model satisfying the desired constraints i.e. $K > 0$, $T_1, \ldots, T_m > 0$, $S_1, \ldots, S_n > 0$. Such constraints appear to be difficult to incorporate into the recursive scheme and a constrained optimisation technique appears to provide the most obvious approach. Having accepted such an approach the modifications to the error criterion to accommodate a linear least-squares solution can be relaxed and the general error criterion E, defined in equation (4), can be utilised with $N(j\omega_k)$ and $D(j\omega_k)$ containing the optimisable parameters $K, T_1, \ldots, T_m$ and $S_1, \ldots, S_m$ respectively. As in the

recursive approach the choice of $w_{k1} = 1/\text{Re}[G(j\omega_k)]$,
$w_{k2} = 1/\text{Im}[G(j\omega_k)]$ has been tested along with the more usual choice
of $w_{k1} = w_{k2} = 1$. To date, a quasi-Newton routine with Lagrange
multiplier checking has been used to perform the constrained optimisation. Results from this latter optimisation approach and the
former RLS technique are presented in the next section.

Another important consideration is the relative efficiency of the
optimisation and RLS procedures. A crucial factor having a direct
bearing on this consideration and one which also has implications for
time domain modelling may now be introduced.

Order of modelling transfer function

The 'best' degree of numerator and denominator polynomials i.e. the
order of the modelling transfer function, is in general unknown. It
is sometimes assumed however that because, say, two time constants
are known to adequately model the time domain response then two
numerator and denominator factors will suffice and that estimating the
transfer function by one of the previously outlined methods will then
yield these dominant time constants. If the frequency response tests
are performed up to relatively high frequencies any faster time
constants, while negligible in the time domain, will affect the
measured data at the higher frequencies. Many distributed parameter
systems are typical of this situation: in theory an infinite number
of time constants characterise the system response though only one or
two may be significant in the time domain. In such situations fitting
a model transfer function of lower order than is detectable in the
given frequency range will not usually yield the dominant time constants.
Sampling up to lower frequencies will eliminate the effect of the
faster time constants but it will also reduce the range of information
from which the slower time constants are to be found. Hence if
frequency response data is given over a wide frequency range, the
fitted model should have enough terms to encompass all significant
time constants in that range even though perhaps only the slowest two
or three time constants will be used in time domain simulations.

The problem of over estimating the number of numerator and denominator
time constants should also be mentioned. As outlined previously, if
the model is chosen with an incorrect difference in the number of
numerator and denominator time constants then an appropriate number of
'negligible' time constants are predicted. If such time constants are
neglected but over estimation of the number of time constants persists,
then experience has shown that both the optimisation and RLS approaches
lead to pole/zero cancellation i.e. identical factors are predicted in
the numerator and denominator. Thus by allowing an over estimation of
the polynomial degrees the correct order of model may eventually be
deduced. The number of parameters in a suitable over estimation may
be considerable and while this does not lead to a large increase in

131

computation time for RLS, such times become significant via the
optimisation approach.

Adaption to Improve Efficiency

With no *a priori* knowledge of the correct model order the only way of
accurately identifying all significant time constants in a given
frequency range is to over estimate the order. Such an over estimation would necessarily contain many terms and the computation time
in applying the constrained optimisation approach would be appreciabl
A more sophisticated scheme utilising the optimisation technique at
its kernel was therefore developed.

By setting a model order, performing the appropriate optimisation and
checking for 'negligible' time constants and pole/zero cancellation,
it is easy to discover whether or not the correct order has been
chosen. If incorrect, the order can be incremented and the latter
procedure repeated. However, as suggested earlier, an under estimate
model to the whole data set will not necessarily give the correct
dominant parameter values and hence such values may not provide good
initial estimates for an ensuing optimisation of parameters in a large
order model. The adaption of the optimisation approach involves
partitioning the whole data set into several individual sets each
containing, say, one additional decade of frequency responses. The
'best' model is fitted to the first data set and the final parameter
values are used as initial estimates for the dominant parameter values
of the second data set. The 'best' model is then found for this large
data set. This procedure is repeated until the whole data set has
been encompassed.

5.   RESULTS

The prediction of suitable transfer functions in synchronous machines
is one aim of this work and as a consequence the results given in this
section are based on the estimation of the parameters K, T_1, $\ldots$, T_m,
S_1, $\ldots$, S_n of form III.

RLS approaches

A fundamental test of the algorithms presented in sections 2 and 3 is
to see how well they estimate the transfer function of uncorrupted
data generated by a specified transfer function. Thus

$$G(s) = 3.0 \frac{(1 + 5.0s)(1 + 0.5s)(1 + 0.05s)(1 + 0.005s)}{(1 + 1.0s)(1 + 0.1s)(1 + 0.01s)(1 + 0.001s)}$$

was used to generate responses at 40 logarithmically equally spaced
frequencies in the range 0.01 to 100 rad/sec. The RLS procedures
were then applied to determine the appropriate parameters of form I
and suitable factorisation then yielded the required parameters K,
T_i $(i = 1, \ldots, m)$, S_i $(i = 1, \ldots, n)$. Table 1 illustrates the effect

		Levy (K)		Lawrence and Rogers (K)		Method 1 (K)	
m	n	T_i	S_i	T_i	S_i	T_i	S_i
2		(6.833)		(4.780)		(3.179)	
	2	0.8508	0.0419	1.523	0.0765	3.904	0.2664
		0.0213	0.0032	0.0336	0.0039	0.0843	0.0067
3		(4.542)		(3.948)		(3.006)	
	3	1.874	0.1230	2.310	0.1522	4.944	0.9293
		0.0548	0.0105	0.0668	0.0140	0.4493	0.0826
		0.0054	0.0011	0.0083	0.0019	0.0383	0.0041
4		(3.000)		(3.000)		(3.000)	
	4	5.000	1.000	5.000	1.000	5.000	1.000
		0.5000	0.1000	0.5000	0.1000	0.5000	0.1000
		0.0500	0.0100	0.0500	0.0100	0.0500	0.0100
		0.0050	0.0010	0.0050	0.0010	0.0050	0.0010
5		(3.000)		(3.000)		(3.000)	
	5	5.000	1.000	5.000	1.000	5.000	1.000
		0.5000	0.2457	0.5000	0.2457	0.5000	0.2458
		0.2457	0.1000	0.2457	0.1000	0.2458	0.1000
		0.0500	0.0100	0.0500	0.0100	0.0500	0.0100
		0.0050	0.0010	0.0050	0.0010	0.0050	0.0010

Table 1. RLS approaches and model order estimation

of under and over estimation of model order. The technique denoted 'Method 1' is the RLS implementation of the relatively weighted error criterion introduced in section 3, while that denoted 'Lawrence and Rogers' is the RLS equivalent of the method of Sanathanan and Koerner Clearly, all methods provide inaccurate dominant parameter values wher the model order is an under estimate of the true order, though the table also shows that Method 1 provides the most realistic values; the frequency response fit is also correspondingly better. When the model order is correct all methods give identical results. The accuracy of the Levy technique in this situation is a little mis-leading: the method only works because the data is exact. The pole/zero cancellation for model order over estimation with no discrepancy in the difference of numerator and denominator polynomial degrees is also evident in all the approaches.

Table 2 shows an improved estimation of the dominant parameters by 'Method 2'; the higher order model results are identical to those in Table 2. This method uses the same error criterion as Method 1 but does not perform any 'outer loop' iterations. These iterations are replaced by an incremental structure: the responses from a certain number of the lowest frequencies are chosen to form an initial data set to which the RLS procedure is applied. The data set is then incremented by including the next frequency data point and the RLS procedure is applied to the new data set using the parameter values from the previous step as initial estimates. This scheme is repeated until the full original data set has been covered.

	Method 2	
m	(K)	
n	T_i	S_i
	(3.063)	
2	4.514	0.4495
2	0.1520	0.0111
	(3.000)	
3	4.990	0.9812
3	0.4846	0.0919
	0.0432	0.0045

Table 2. Improved dominant parameter estimation

Both the method of Lawrence and Rogers and Method 1 perform 'outer loop' iteration until convergence of the model parameters is noted. Table 3 illustrates an example of incorrectly selecting the difference in the degrees of numerator and denominator polynomials

	Iteration 5		Iteration 6	
m	(K)		(K)	
n	T_i	S_i	T_i	S_i
	(3.000)		(3.000)	
4	0.5000E+01	0.1008E-02	0.5000E+01	0.1004E-02
4	0.5000E+00	0.1000E-01	0.5000E+00	0.1000E-01
	0.8725E-04	0.1000E+00	0.6815E-04	0.1000E+00
	-0.7954E-04	0.1000E+01	-0.6383E-04	0.1000E+01

Table 3. Model order over estimation

and its effect on convergence. The data was generated by the transfer
function

$$G(s) = 3.0 \; \frac{(1 + 5.0s)(1 + 0.5s)}{(1 + 1.0s)(1 + 0.1s)(1 + 0.01s)(1 + 0.001s)}$$

at the 40 frequencies previously used, fourth order polynomials being
selected in the model numerator and denominator. The convergence of
the dominant parameters is clear, though the presence, and lack of
convergence, of 'negligible' parameters is also evident.

The basic RLS methods have also been applied to several experimental
data sets. Figures 1 and 2 show the fits achieved by fixed order
models (m = 4, n = 3) via the methods of Lawrence and Rogers and
Method 1 respectively. Levy's method was also applied but it gave a
much poorer fit and determined certain of the time constants to be
negative. The data is from an NEI Parsons EPRI report (1980),
table B-1, and was originally derived from standstill frequency
response tests to determine an impedance function Z(s) for a large
turbogenerator.

Constrained optimisation approach
The procedure suggested in section 4 has been implemented and appears
to be stable in its operation on both deterministic and experimental
data. Figure 3 shows the result of applying the scheme to experimental data as presented in the NEI Parsons EPRI report, table B-2.
As shown a high order model is determined by the procedure. This
model accurately characterises the data and should also adequately
predict the dominant time constants.

Computation times
All the CPU times which follow are for the specified computation
performed on a PRIME 750 computer.

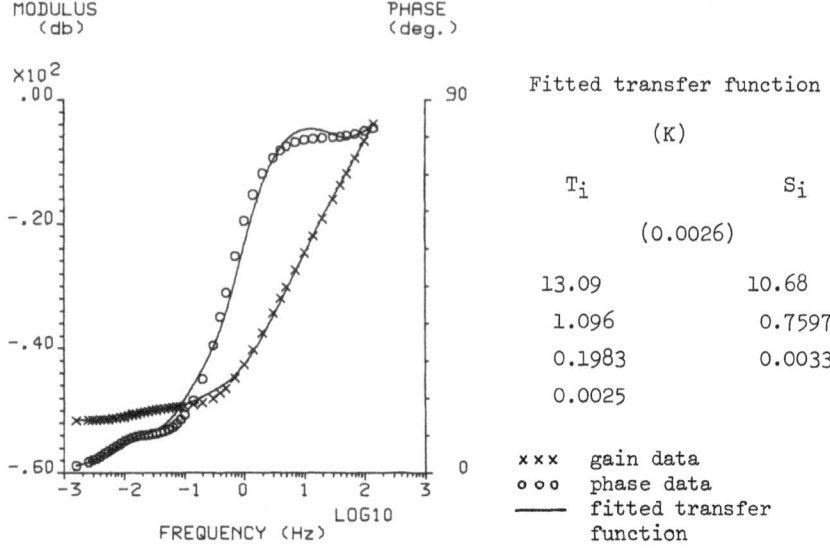

Figure 1. Lawrence and Rogers method

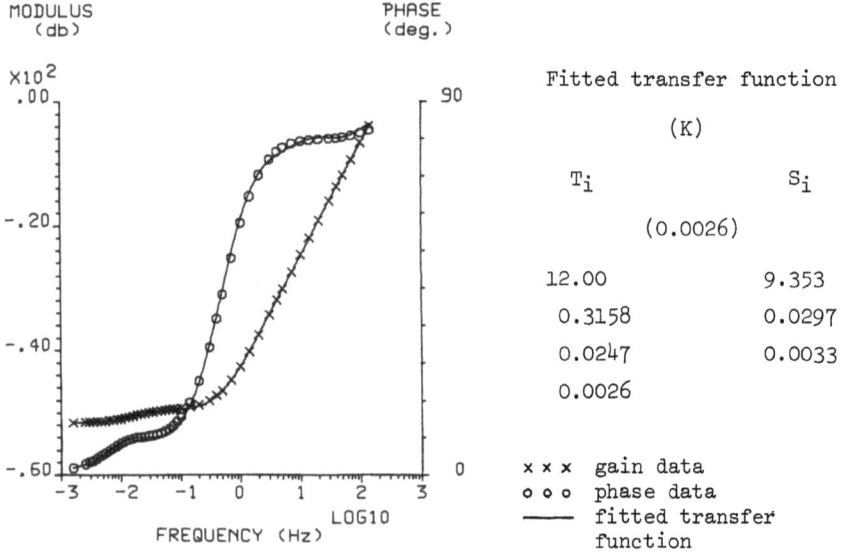

Figure 2. Method 1

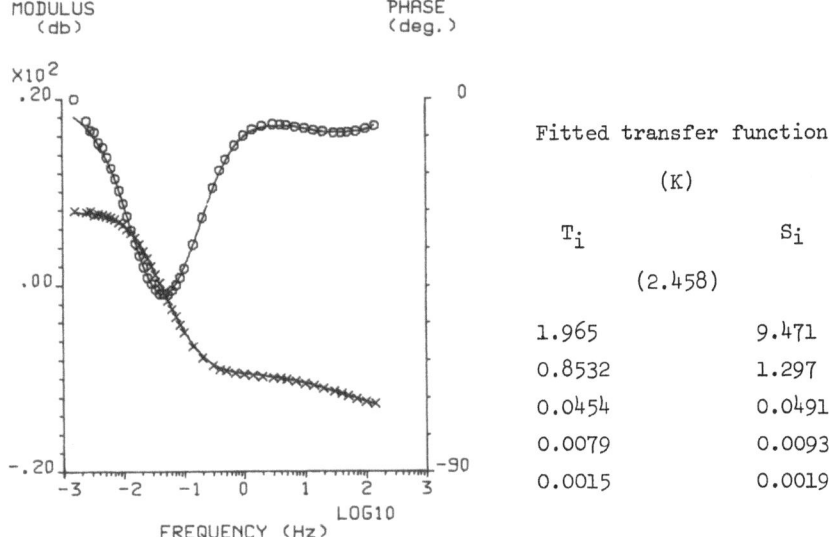

MODULUS
(db)

PHASE
(deg.)

Fitted transfer function

(K)

T_i	S_i
(2.458)	
1.965	9.471
0.8532	1.297
0.0454	0.0491
0.0079	0.0093
0.0015	0.0019

Figure 3. Constrained optimisation

Levy's method involves computation for the order of 1 second while
Lawrence and Rogers' method and Method 1 take approximately 3 seconds
for deterministic data but around 40 seconds for experimental data
($m = n = 5$). The constrained optimisation approach computes for
approximately 40 seconds for deterministic data and anything up to
600 seconds for experimental data, though this scheme does include
automatic determination of 'best' model order.

6.   CONCLUSION

The restructuring of the problem of estimating transfer functions
of general form I via a recursive least-squares solution has led to
this study wherein several alternative methods of prediction
may be easily evaluated. Levy's method is not recommended for use on
experimental data. While the method of Lawrence and Rogers
(Sanathanan and Koerner via RLS) gives a reasonable fit to experimental
data the technique based on the relatively weighted error criterion
(Method 1) usually proves to be superior.

When employing the latter techniques to determine transfer functions
of form III with positive parameters, models of the required

structure are generally predicted. However this cannot be guaranteed and failures of both methods have been noted. The constrained optimisation approach outlined in section 4 offers the most robust solution to this problem.

The nature of selecting a model of incorrect order has been presented in the paper. It is emphasized that if the purpose of determining the transfer function is to predict dominant time constants then either an appropriate and limited range of frequencies for the number of desired time constants should be selected, or the model order should be chosen to encompass all possible time constants in the given frequency responses. Pole/zero cancellation and/or 'negligible' time constants are positive indicators of model order over estimation and these factors appear to be the key to finding the correct model order.

REFERENCES

Lawrence, P. J. and Rogers, G. J., "Transfer function synthesis from measured data", Proc. IEE, Vol. 126, No. 1, pp. 104-106 (1979)

Levy, E. C., "Complex curve fitting", IRE Trans. on Automatic Control, Vol. AC-4, pp. 37-43 (1959)

NEI Parsons Ltd, "Determination of synchronous machine stability study constants", EPRI EL-1424, Vol. 4 (1980)

Payne, P. A., "An improved technique for transfer function synthesis from frequency response data", IEEE Trans. on Automatic Control, Vol. AC-15, pp. 480-483 (1970)

Sanathanan, C. K. and Koerner, J., "Transfer function synthesis as a ratio of two complex polynomials", IEEE Trans. on Automatic Control, Vol. AC-8, pp. 56-58 (1963)

Young, P. C., "Recursive approaches to time series analysis", IMA Bulletin, pp. 209-224 (1974)

APPENDIX A   RECURSIVE LEAST-SQUARES ALGORITHM

Consider the problem

$$\min_{x_1, \ldots, x_q} \sum_{i=1}^{p} \left(\sum_{j=1}^{q} c_{ij} x_j - d_i \right)^2$$

i.e. $$\min_{\underline{x}} \sum_{i=1}^{p} (\underline{c}_i^T \underline{x} - d_i)^2$$

138

Provided the matrix $c_i c_i^T$ is non-singular, the normal equations give the minimising vector, $\hat{x}_p$, as

$$\hat{x}_p = P_p \, h_p$$

where
$$P_p = \left[\sum_{i=1}^{p} c_i c_i^T \right]^{-1} ; \quad h_p = \sum_{i=1}^{p} c_i d_i$$

An alternative solution (RLS) is to compute $\hat{x}_p$ and P_p recursively from

$$\hat{x}_p = \hat{x}_{p-1} - P_p [c_p c_p^T \hat{x}_{p-1} - c_p d_p]$$

$$P_p = P_{p-1} - P_{p-1} c_p [1 + c_p^T P_{p-1} c_p]^{-1} c_p^T P_{p-1}$$

$\hat{x}_0$ is set as arbitrary and P_0 as diagonal with elements of order 10^6; this comes from a statistical consideration of the problem wherein P_p is identified as a variance-covariance matrix and the values in P_0 reflect the lack of confidence in the choice for $\hat{x}_0$. Thus this algorithm provides an updated estimate of the unknown parameter vector on receipt of the lastest i.e. p-th "data set" $c_{p1}, c_{p2}, \ldots, c_{pq}, d_p$.

APPENDIX B   RLS IMPLEMENTATION OF E_r^*

For the general error criterion E_r^* posed in Equation (6), the p-th data set (p = 1, 2, ..., 2M) required by the RLS algorithm is given as follows.

For p odd (real part contribution)

$$
\begin{aligned}
c_{pj} &= (-1)^{(j-1)/2} \, \omega_k^{(j-1)} \, w_{rkl} \qquad && \left. \begin{array}{l} j \text{ odd} \\[4pt] j \text{ even} \end{array} \right\} \; j = 1, \ldots, m+1 \\[2pt]
&= 0
\end{aligned}
$$

$$
\begin{aligned}
c_{pj} &= \beta_k (-1)^{(j-m-1)/2} \, \omega_k^{(j-m)} \, w_{rkl} \qquad && \left. \begin{array}{l} (j-m) \text{ odd} \\[4pt] (j-m) \text{ even} \end{array} \right\} j = m+2, .., n+m+1 \\[2pt]
&= \alpha_k (-1)^{(j-m+2)/2} \, \omega_k^{(j-m)} \, w_{rkl}
\end{aligned}
$$

$$d_p \doteq \alpha_k \, w_{rkl}$$

For p even (imaginary part contribution)

$$c_{pj} = 0 \qquad\qquad\qquad\qquad\qquad\qquad j \text{ odd}$$
$$= (-1)^{(j+2)/2} \, \omega_k^{(j-1)} \, w_{rk2} \qquad j \text{ even} \Bigg\} j = 1, \ldots, m+1$$

$$c_{pj} = \alpha_k (-1)^{(j-m+1)/2} \, \omega_k^{(j-m)} \, w_{rk2} \qquad (j-m) \text{ odd}$$
$$= \beta_k (-1)^{(j-m+2)/2} \, \omega_k^{(j-m)} \, w_{rk2} \qquad (j-m) \text{ even} \Bigg\} j = m+2, \ldots, n+m+1$$

$$d_p = \beta_k \, w_{rk2}$$

where

$$k = \text{integer part of } (p + 1)/2$$
$$\alpha_k = \text{Re}[G(j\omega_k)]$$
$$\beta_k = \text{Im}[G(j\omega_k)]$$

and the other vector of unknowns is ordered $(a_0, a_1, \ldots, a_m, b_1, \ldots, b_n)$.

SESSION E

SIMULATION & COMPUTING TECHNIQUES II

Chairman

M J O'CARROLL

Teesside Polytechnic

THE EXTRACTION OF ENGINEERING INFORMATION FROM A POTENTIAL SOLUTION TO AN ELECTROMAGNETIC FIELD PROBLEM

D.A.Lowther

Computational Analysis and Design Laboratory
McGill University.
Montreal.
Canada.

1. INTRODUCTION

In recent years it has become necessary, for both technical and economic reasons, to obtain accurate solutions for the electromagnetic field in complex devices. The complexity may be introduced by the geometric shape, the material properties, time dependence or a combination of all of these. Traditionally, these devices have been designed using a mixture of simplified, algebraic, closed form solutions, experimental evidence and design experience. However, the cost of prototypes may be prohibitively high, there may be little or no experience available, and algebraic methods, whilst convenient to use, are unable to model many of the complexities.

The digital computer has provided the designer with an alternate tool which may be used as part of a design process to develop solutions to some of these problems and, although a full three-dimensional solution with non-linearity and time-dependence is still not a routine computation, two-dimensional field calculations have become commonplace. These systems are almost all based on some form of piecewise (discrete) approximation to the field, i.e. a numerical method, rather than a computer implementation of an algebraic solution. The advantages of numerical methods lie both in the ability to deal with complexity by reducing the problem to a large number of simple pieces and in the flexibility afforded by the ability to use a single computer program for a wide range of problems.

Several approaches to the numerical solution of a field problem exist and these can be divided into two distinct classes -- differential and integral. The choice of approach is influenced by several factors, of which the main ones are the requirement for data only at specified positions or globally, and the available computer. With current computing architectures the differential approach is the most favoured, although developments in parallelism may well alter this situation. In this paper it is assumed that the solution method is differential and based on the finite element technique.

If a finite element approach is used, the solution is performed in

143

terms of a potential function and a best fit to the true field solution is obtained by approximating the field piecewise by a set of polynomials. The resultant solution, in the computer memory, is a set of potentials at the nodes of the elements. Information about the polynomials used over each element is implicit in the solution technique. Thus the stored data is, in fact, a ´tangible´ model of the real field. However, it is, in itself, of little use to the design engineer and further processing is required to extract information which is relevant to the design. This secondary processing is the main subject of this paper.

2. PROBLEM SPECIFICATION AND SOLUTION

To obtain a computer solution to a design problem several stages are necessary (Silvester, 1981). Firstly, the designer has to describe the problem to the computer. This might be considered to be a process by which an engineering problem is converted into a mathematical problem. This process may be further subdivided into two parts:

a) The geometric description and
b) The properties and boundary conditions description.

Each of these tasks requires an input system which accepts engineering data from the designer. The two parts are combined to produce a mathematical description of the problem.

The second phase involves the construction of a set of equations, based on a particular potential function, which describes the system. At the end of this phase the set of equations is solved to produce data in the form of potential values which describe the field. This data, however, is purely mathematical in form and is often of little direct use to the designer.

In the course of a design, the engineer may well require calculations of impedances, forces, losses, harmonic content of waveforms, etc. These quantities may all be derived from the mathematical solution, but require further processing. This process has become known as the postprocessing, or result evaluation phase, of a CAD system.

3. RESULT EVALUATION

Unlike the preprocessing phase, it is difficult to predict in other than broad terms the information that might be needed as a final output from a postprocessor. Typically the output might include flux plots, energy and inductance calculations, and terminal parameter evaluation, as might be expected from a traditional design system.

144

These might be considered as ´global´ parameters. At a more ´local´
level there may be a requirement for loss calculations, local field
values, etc. Beyond these, however, more specific data may be
required such as the actual field distribution over the pole face of
a machine in order to locate potential hot spots or to detect
possible commutation problems.

Operations, in addition to calculation facilities, might include the
creation and recording of graphs, plots, and numeric data, thus
requiring both a graphics display and some form of mass storage.

The following sections consider typical operations of the global,
local and distributed types.

Global Parameters

These are parameters which depend on the overall performance of the
device, i.e. local effects are ´smoothed out´. Typically, the
determination of the terminal parameters of a magnetic device is
usually among the most important if it is to be used as part of a
mechanical or electrical circuit. These include both the impedances
and the mechanical output. Although the potential solution does not
provide this information directly, it may be derived from the energy
contained in the system and the energy changes which occur as
relative movement occurs. Thus the calculation of the energy stored
within the device is important.

The finite element representation of Poisson´s equation is given by

$$S A = - T J \tag{1}$$

where the matrices S and T represent the finite element matrices
associated with the boundary value problem that relates the unknown
potentials A to the known sources J. The energy W within any one
element e is given by

$$W = \frac{1}{2} A \, dS \tag{2}$$

or

$$W = \frac{1}{2} A S A \tag{3}$$

Thus the total energy within the model is given by:

$$W = \frac{1}{2} A S A \tag{4}$$

where A represents the column vector of all the nodal potentials.

145

Also, from Equation (1), the energy is given by

$$W = \frac{1}{2} A T J \qquad (5)$$

Equation (4) represents the stored magnetic energy within the system, whereas Equation (5) gives the energy input from the source. It should be remembered that the magneto-static system has no dissipation elements and thus cannot represent the ohmic losses in, for example, the windings of a machine. Using Equations (4) and (5), the energy distribution within the machine and the total energy, can be found. If the system is non-linear care must be taken in using these equations because the energy stored in any element is dependent on the area of the B-H curve.

As an example of a terminal parameter calculation, the inductance of the device may now be found from the stored energy by using the relationship

$$W = \frac{1}{2} L I \qquad (6)$$

where I is the input current to the device, i.e. is given by

$$I = J \, dS \qquad (7)$$

An alternative approach would be to consider the concept of flux linkage; this could be required for both inductance and voltage calculations. The calculation of flux linkage is described later.

Local Field Values and Components - Points, Lines and Zones
The advantage of a numerical solution is in the detailed description of the field throughout the device. The data available can be compared to that which might be obtained from an experimental rig; instrumentation can be provided to measure the field at selected points within the device. Similarly a designer can be provided with methods of determining the field value at a point by specifying the appropriate point. Once the element containing the point has been determined the value of the potential can be found from a knowledge of the polynomials used within the element. Thus the analytical data may be provided in a form similar to that obtained from experiment.

Often it is desired to have not only point values of the field, and its components but also its spatial distribution over a part of the device. For example, the radial component of the air gap flux density in an electrical machine gives both a qualitative and quantitative indication of the rotor/stator coupling and may also be used in the calculation of the forces exerted on the shaft. The calculation of the force exerted on a magnetic or current carrying body in an electromagnetic field can be calculated in several ways; either the cross product of current density and flux density integrated over a surface, or by a Maxwell stress formulation over a contour enclosing the body (Carpenter, 1960). To provide calculations of this form the designer should be able to specify a contour through the device along which he wants to calculate the field or one of its components.

Some information may not be calculable over either a point or contour and requires the specification of a zone within the device. To calculate terminal voltage of a winding, for example, in the postprocessor it must be possible to specify an area over which the calculation is to be performed, in this case the conductor cross-section, and to perform both a dot product, integration and division operations over this area.

Consequently, to use the detailed data provided by a numeric analysis to improve the design, the designer must be able to specify sections of the device for examination. These sections could range from a complete area within the model, such as might be needed in the calculation of the loss in a tooth, to a one-dimensional piece, i.e. a contour, along which the distribution of the field is required, to a zero-dimensional piece, i.e. a point, at which the local value of the field, or its value in a given direction might be needed.

Integration of Distributed Values - Flux linkages and Losses

Often, the set of values obtained locally is useful both as a graphical distribution and as the basis for an integral, such as the loss in a pole face, or the flux per pole. Each of these may require careful consideration of the potential used.

For example, the flux linkage of a single turn coil of cross-section $´a´$ and length $´1´$ is given by:

$$\Phi = Ba = \iint \underline{B}.d\underline{S} \tag{8}$$

Substituting for the flux density, B, with the vector potential A,

$$\Phi = \int \underline{A}.d\underline{\ell} = A\ell \tag{9}$$

However, the vector potential is not necessarily uniform over a complete conductor of finite cross-section; indeed, the conductor may be made up of several bars. Thus the value of A in the above equation should be replaced by a weighted average over the conductor:

$$A\ell = \Sigma \underline{A}.d\underline{\ell} \tag{10}$$

Consequently the flux per pole can be determined without recourse to

147

numerical differentiation and its associated problems.

In many designs a magnetostatic solution is sufficient to provide the required results, especially' if the device has no paths in which currents may be induced. However, if there is relative motion of two or more sections of the device and this is periodic, as in an electric machine, then each part of the device is subjected to a field which is constantly changing. This will, in turn, result in losses within the material. These losses are often described in terms of the magnetic flux being carried by the material and the frequency of the flux waveform. The results are supplied as a set of loss curves, which should be available to the result evaluation system. The loss may then be determined by accessing the loss curve with the local value of flux.

4. GENERAL-PURPOSE INTERACTIVE POSTPROCESSORS

Although postprocessing either by hand or built into the solver system has been performed routinely, two main approaches have developed for interactive result evaluation. The first allows the user to specify the output required before the solving phase occurs -- the output of the solver is then the postprocessed data. This is a limited form of interaction but very suitable if the solution process is running on a remote, central computer. The second approach, and that to be discussed here, allows the user to manipulate the output from the solver after completion of the solver phase.

Postprocessing Principles
The open-endedness of this form of result evaluation makes the postprocessor substantially different in both its construction and operation from either the data input or solution phases. The user should be allowed to manipulate the data output by the solver in order to derive useful results. In doing so it should not provide operations which are invalid because of assumptions made earlier in the design process, and should not introduce features which did not exist in the original data.

The system philosophy is analogous to that of a hand calculator. The designer of the calculator does not know the end use to which the machine may be put, but the properties of the input data, i.e. single numbers, are understood and a valid set of operations may be defined, each operation being controlled by one key. Similarly for field result evaluation, the properties of the solution potential are known and a valid set of operations may be defined. Again, the hand calculator is an interactive system – a key is pressed and the answer appears almost immediately. This speed is a major reason for the usefulness of the device. The postprocessor should also be capable of providing results quickly and in a format which may be easily understood.

The majority of the results which may be validly derived from the potential solution involve the use of the operations of vector analysis --the operations 'cross', 'dot' and 'curl' have been used frequently in the earlier parts of this paper. One scenario is, then, a postprocessor which provides the user with these operations. However, this approach may well be at too low a level for daily design work and a user may well prefer higher level concepts such as 'flux density', 'impedance' and 'force' to those of 'divergence' and 'curl'. If the high level approach is adopted then the system may become too restrictive and the user becomes limited by the postprocessor designer.

5. MATHEMATICAL PRIMITIVES IN POSTPROCESSING

The field that a postprocessor can expect from the solver is either vector or scalar in nature. The final results that a user requires can be extracted from the solution by the operations of vector analysis. Thus, if a vector potential has been used, the flux density can be obtained by

$$\text{curl } A = B \qquad\qquad\qquad (11)$$

A directional derivative such as B normal to a specified contour would then require the dot product with a unit vector normal to the contour.

The mathematical functions that should be available within the postprocessor are then the following:

Vector Analysis:	curl
	div
	grad
	contour integral
	surface integral
Vector Algebra:	cross-product
	dot-product
	addition
	subtraction
Scalar Algebra:	addition
	subtraction
	multiplication
	division

Additionally, the design engineer may be interested in global concepts such as the input impedance or terminal voltage. If the

solution has been performed in terms of a vector potential the output
voltage calculation algorithm can be constructed in terms of the
potentials themselves, as described earlier, thus avoiding any need
for the differentiation that would be required if it were evaluated
in terms of the flux density.

Equally as important as the mathematical operations provided are the
methods of displaying results, these can be both numeric and
graphical.

6. GRAPHIC DISPLAY AND INTERACTION

In any interactive system, the user should not only be able to
understand what he is being told with a minimum of effort but should
also be able to control the way in which the information is
displayed. The forms of display should, if possible, conform to
current design conventions. These provide two-dimensional and one-
dimensional graphics in the forms of equipotential plots and graphs.
Interactive input, ideally, should be by some form of ´pointing´
device such as a pen and tablet.

Flux and Equipotential Plots
Although numerical results provide a designer with detailed
information about the performance of a magnetic device, considerable
qualitative data is available from a flux plot for a two-dimensional
analysis. The pictorial representation of the field can provide
information both about the correctness of the data provided to the
solution system and the field intensities in various parts of the
device. For preliminary design work this representation of the field
may well be sufficient to indicate possible changes to the design
without any numeric data being needed.

To obtain plots of flux distributions in two-dimensional magnetics it
suffices to draw contours of constant vector potential A (in x-y
problems) or rA (in r-z problems).

Since the postprocessing phase of the design system has access to
both the geometric data base, describing the element configurations,
and the potential solution, and a knowledge of the approximating
polynomials used over each element, the intersection point between a
specified potential value and an element edge can be quickly
determined. Thus, if the edge intersection points are found, and the
variation within the element is known, the equipotential can be
plotted.

Integration Contours and Regions
As described earlier, the ability to examine the data over specified
regions of the device is needed for several purposes, such as
voltage, force and loss calculations. All these operations require

the specification of a section of the data base over which the functions should operate. These subsets of the initial data base may have several forms. The first is the ´zone´ -- a two-dimensional section of the model, the second is the ´contour´ -- a one-dimensional section, i.e. a line, through a section of the model (for example, the air-gap of an electrical machine), and the third is the ´point´ -- a zero-dimensional section of the model. The valid set of operations over each of these geometric entities is, naturally, different -- it is difficult to differentiate a point value!

Graphs and Plots
The previous sections have stressed the need for a method of defining curves or contours within the device. These allow the evaluation of various field quantities along a specified line through the model. The conventional method of presenting data calculated as a function of one variable, in this case the distance along the curve, is a graph. A postprocessing system should allow a user to specify the type and format of the graph to be produced.

A second format for data along a curve might well be tabular, this form is essentially numeric in nature and more complex to understand - a typical curve might have 500 values along its length. Presented as a graph salient features can be spotted quickly, as a set of numbers these are not as easy to find. However, if further processing of the data is required, the tabular form is usable by the computer, provided that it can be saved in an external file.

Calculation over an area requires a different form of display -- the simplest example of this is the flux plot, mentioned earlier. Equipotential plotting can be extended to cover features such as induced current densities, loss densities, etc. An alternative form of display requires grey-scaled or colour graphics. Areas of specific values can be coloured to indicate areas of high and low density. Both these techniques are familiar in other fields of engineering. In general they provide rough qualitative information rather than detailed quantitative data.

7. PROGRAMMABILITY

The calculation functions described above, although providing flexibility in the manipulation of the input data, can make the result evaluation phase of an analysis system difficult and complex to use. An alternative approach would be to define a set of higher level variables which are based on a knowledge of the properties of the approximation method used for the solution. These variables could be the more usual field variables such as magnetic flux density, the magnetic field intensity, the permeability, etc. This not only presents the user with a set of variables with which he is familiar but also reduces the number of functions that have to be

151

provided by the processor, thus simplifying its design, and possibly decreasing its response time.

Although the latter approach provides a simple system for the novice or routine design engineer, it is probably not suited to a research environment in which several problem formulations may need to be considered. In this case, although the ´high level´functions may have a meaning, it can become confusing to use them. Thus, a better approach would be to use a processor built only with a knowledge of vector analysis and the properties of finite element solutions.

One possible solution to this problem is to provide a level of programmability within the processor. Thus commonly used sequences of operations may be stored, providing the appearance of the simple processor. For example a command ´FLUX´may be defined which will calculate the flux density by executing the CURL, DOT and MAGNITUDE functions of the underlying processing system.

Each set of stored operations becomes a new command to the system, allowing a user to configure the processor to his preferences and to make it easy for his application. The approach is similar to that provided in many modern operating systems.

By allowing parameter substitution at execution time, user defined commands can appear to have all the capabilities of the basic system commands and can use qualifiers which are also user defined. The particular set of commands used by one user may not be the same as those of another user, so that each user requires a profile file which contains, not only his own command set but also information about default values for graphing parameters, colour maps, interpolation polynomials, etc.

8. IMPLEMENTATION EXAMPLES

At least two general purpose, electromagnetic field postprocessors of the type described are in reasonably wide-spread use. These are the postprocessor for the Magnet Eleven system (Lowther, 1982) and the Ruthless post-processor (Lowther, 1981). The latter is an example of the flexible type of system and was designed to interface to very large finite element programs. It provides the user with the basic vector operations and relatively sophisticated graphics displays. The system runs on a PRIME or VAX computer and thus has access to very large virtual memory spaces. The bare, low level vector analysis route was chosen in this case because the user community was not likely to be design engineers accustomed to conventional design methods, but rather research engineers and scientists examining more fundamental aspects of the solution process itself.

The MagNet processor was implemented on a micro-computer system and has considerably less flexibility. The variables it can manipulate and the operations provided are much closer to those found in conventional design. The computer system is low enough cost that each engineer may have his own, and efficient resource sharing is no longer a major factor. More recently, the MagNet processor has been developed towards the Ruthless approach but maintains the original hardware configuration.

7. CONCLUSIONS

A postprocessor is a key part in any design system which includes an analysis phase. It allows relevant data to be extracted from the solution and presented in a way which is both of meaning to, and controlled by, the user. To be effective it should be an interactive process, providing the designer with answers to questions about the solution relatively rapidly. Ideally the implementation should be on a computer which is sufficiently available that an engineer can have constant access to the data. Such a system may well be micro-processor based and in this situation considerable care must be taken in the basic design of the system to achieve the necessary response.

The increasing power and decreasing cost of small machines will mean that postprocessing techniques such as those described above will become commonplace in the design and research offices.

10. REFERENCES

Carpenter,C.J., "Surface-integral methods of calculating forces on magnetised iron parts", Proc. IEE, 107C, 19-28, (1960).

Lowther, D.A., Silvester, P.P., Freeman, E.M., Rea, K., Trowbridge, C.W., Newman, M. and Simkin, J., "Ruthless - a general purpose finite element post-processor", IEEE Trans., MAG-17, 3402-3404, (1981).

Lowther,D.A., "A microprocessor based electromagnetic field analysis system", IEEE Trans., MAG-18, 351-356, (1982).

Silvester, P.P., Lowther, D.A., Freeman, E.M. and Csendes, Z.J., "The preprocessing, solution and postprocessing of finite element electromagnetic field problems using a small dedicated computer with raster graphics", EngSoft.Conf., Imperial College, London, (May 1981).

OPTIMISATION OF THE MAGNETIC SCREENING OF ELECTROMAGNETIC COILS

A Zisserman[1], J Caldwell[2] and D H Prothero[3]

1. Department of Mathematics and Computer Studies,
   Sunderland Polytechnic.

2. School of Mathematics, Statistics and Computing,
   Newcastle upon Tyne Polytechnic.

3. International Research and Development Company,
   Newcastle upon Tyne.

1. INTRODUCTION

A frequent problem with systems incorporating coil windings or permanent magnets is the need to reduce the stray magnetic field. This may
be necessary to ensure the operation of nearby instruments sensitive
to magnetic field. The usual solution is to shield the system using
a shell of ferromagnetic material. For small scale applications, a
high permeability alloy, such as mu-metal would be used. However for
large scale applications, mu-metal is not attractive, due to its high
cost, and a material with a characteristic similar to that of mild
steel would be considered.

The present paper examines quantitatively the effect of magnetic
screening on a system of axially symmetric coils. Fig 1 represents
the coils and the screen surrounding them. The purpose of the paper
is to describe a method which can be used to assess the magnetic
screening for a wide range of axisymmetric systems similar to Fig 1.

For the present work, two important simplifying assumptions have been
made:

(i) The coil system shown in Fig 1 is assumed to be surrounded by
    a spherical shell of screening material. The geometry was
    chosen because it is a convenient one for the method of analysis
    used in this paper. In practice, a purely spherical shell
    could not be used; however the results given here will indicate
    the performance of a shield of similar geometry.

(ii) A full anaylsis of the behaviour of a ferromagnetic shield is a
     complex problem due to the variation of magnetic permeability of
     the screen with the magnetic field $\underline{H}$. This would therefore re-
     quire a computer program using numerical techniques. In this
     paper, a simpler, though more approximate, approach is adopted.

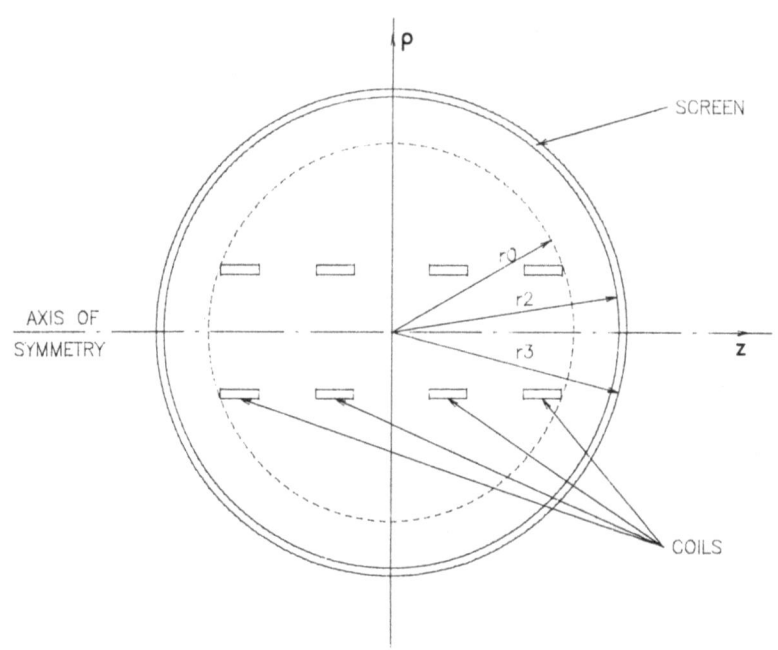

FIGURE 1.   GEOMETRY OF THE SYSTEM

In Section 2 of this paper the effect of a spherical iron shell on the field system is described. Using permeability data for available materials, the actual permeability that the shell adopts, and its variation with angle and shell radius, is described in Section 3. This information is then used to determine the optimum screen radius. Finally in Section 4 an assessment is given of the method which has been used and the results which have been obtained.

2. THEORY

Calculation of Magnetic Field

In the absence of iron the field $\underline{H}$ may be derived from a scalar potential V which satisfies Laplace's equation $\nabla^2 V = 0$. The potential can be expressed as a Legendre polynomial expansion

$$V = -\sum_{n=1}^{\infty} \frac{B_n P_n (\cos\theta)}{(n+1) \, r^{n+1}}$$

where (r,θ) are spherical polar coordinates and $\{B_n\}$ is set of co-efficients determined by the coil geometry and current density. (See Garrett, 1962).

The magnetic intensity $\underline{H}$ is given by $-\nabla V$. So, for example

$$H_r = \sum_{n=1}^{\infty} \frac{B_n P_n (\cos\theta)}{r^{n+2}}$$

The series converges for r greater than r_o, the outermost radius of the coils. (See Fig 1).

For convenience and greater generality the normalised distances (R, P, Z ...) are used throughout this paper, instead of the actual distances (r, ρ, z ...).

Table I compares the calculation of the field using the Elliptic Integral method (Culwick, 1965) with values obtained using the Legendre polynomial expansion. The series converges rapidly especially for larger R ie. $R > R_o$. For speed of computation the series is truncated at n = 30; this gives an error only in the third significant figure for R > 3 and only in the fifth significant figure for R > 4.
Unless otherwise stated the units used throughout the paper are H/Am^{-1} and θ/rad.

Effect of Iron Shell

The effect of the spherical iron shell is included by using the

Table I : Comparison of values obtained from the Legendre Polynomial
expansion (LP) and from the Elliptic Integral (EI) program.

P	Z	B-Radial /KG		B-Vertical/KG	
		EI	LP	EI	LP
0	2.5	0	0	-1.2223 E+1	1.4013 E+1
0	3.0	0	0	-4.6525	4.6571
0	3.5	0	0	-1.6742	1.6746
0	4.0	0	0	-6.7776 E-1	6.7797 E-1
0	4.5	0	0	-3.0910 E-1	3.0922 E-1
0	5.0	0	0	-1.5542 E-1	1.5548 E-1
5	0	-2.7174 E-2	2.7189 E-2	1.0681 E-6	0
7	0	-3.6387 E-3	3.6405 E-3	-1.5259 E-8	0
9	0	-7.0676 E-4	7.0717 E-4	9.5367 E-10	0
11	0	-1.6290 E-4	1.6300 E-4	2.7657 E-7	0
13	0	-3.6182 E-5	3.6189 E-5	2.2289 E-8	0
15	0	-3.2046 E-6	3.2177 E-6	0.0	0
R	θ/rad				
5	1.1	2.5489 E-2	-2.5504 E-2	-1.8849 E-2	1.8860 E-2
7	1.1	4.0222 E-3	-4.0248 E-3	-1.9070 E-3	1.9081 E-3
9	1.1	9.4812 E-4	-9.4876 E-4	-2.7171 E-4	2.7195 E-4
11	1.1	2.9226 E-4	-2.9277 E-4	-2.5441 E-5	2.6170 E-5
13	1.1	1.0837 E-4	-1.0832 E-4	1.3817 E-5	-1.3944 E-5
15	1.1	4.6287 E-5	-4.6212 E-5	1.7343 E-5	-1.7298 E-5

following expressions for the potential:

Region I: $R \leq R_2$ $\quad V = \sum\limits_{n=1}^{\infty} (\dfrac{F_{1n}R^n}{n} - \dfrac{1}{(n+1)R^{n+1}}) B_n P_n$

Region II: $R_2 < R < R_3$

$$V = \sum\limits_{n=1}^{\infty} (\dfrac{F_{2n}^A R^n}{n} - \dfrac{F_{2n}^B}{(n+1)R^{n+1}}) B_n P_n$$

Region III: $R \geq R_3$ $\quad V = \sum\limits_{n=1}^{\infty} - \dfrac{F_{3n} B_n P_n}{(n+1)R^{n+1}}$

where $\quad F_{3n} = \mu (2n+1)^2 / \{ (\mu n + \mu + n)(n+1+\mu n)$

$$- (\mu - 1)^2 n(n+1) (R_2/R_3)^{2n+1} \}$$

$$F_{2n}^A = - F_{3n} n (\mu - 1) / \{ \mu (2n+1) R_3^{2n+1} \}$$

$$F_{2n}^B = F_{3n} (n + n\mu + 1) / \{ \mu (2n+1) \}$$

$$F_{1n} = \mu F_{2n}^A + (\mu F_{2n}^B - 1)/R_2^{2n+1}$$

R_2, R_3 and μ are, respectively, the normalised inner and outer radii of the shell and its constant permeability.

These expressions are exact provided the iron shell has constant permeability and are approximately true if μ is fairly constant. However, if μ does vary within the shell then

$$\nabla^2 V = - \underline{\nabla} V . \underline{\nabla} \mu = \underline{H} . \underline{\nabla} \mu$$

Determination of permeability $\mu (\theta, R_2)$

The permeability of the shell will depend on the magnitude 'H' of the magnetic intensity $\underline{H}$ inside the shell ($H = |\underline{H}|$) and H in turn depends on the permeability 'μ' of the shell. For the purpose of the calculation the mid-shell (ie. at $R = (R_2+R_3)/2$) value of H is used. The actual permeability adopted by the screen will be such that:

$$\mu(\bar{H}) = \bar{\mu} \quad \text{where} \quad H(\bar{\mu}) = \bar{H}. \tag{1}$$

159

Also H is a function of two variables θ and R_2, and so the same applies
to μ. In order to determine $\mu(\theta, R_2)$ satisfying condition (1) μ as a
function of H and the family of curves $H(\theta, R_2)$ are needed.

In Fig 5 the curves are shown for a cubic spline fit to the permeability data of two materials. Data 1 corresponds to mild steel and
Data 2 corresponds to a high alloy steel.

3. RESULTS

Calculation of permeability and magnetic field values

By way of illustration, the results for the case $R_2 = 3.0$ are described
in some detail.

The behaviour of H_r and H_θ inside the screen (at $(R_2 + R_3)/2$) is shown
in Figs 3 and 4 for the constant permeability $\mu = 10$. H_r is now the
smaller component, since it is reduced by a factor of μ^r relative to
its value outside the screen, and so the behaviour of H shown in Fig 5
is dominated by H_θ. As the permeability is increased the curve $H(\theta)$
has the same shape but is reduced in magnitude.

The curves $\mu(H)$ and $F(\theta)$ are used to determine $\mu(\theta)$. $\text{Log}_{10}(\mu(H))$ and
$\text{Log}_{10}(H(\mu))$ are plotted in Fig 6 and the intersections give $\mu(\theta)$.
The curves of $H(\mu)$ for different values of θ do not cross showing that
in this range of μ the shape of the $H(\mu)$ curve is maintained. The
coordinates of the intersections are determined iteratively. The
shape of the $\mu(\theta)$ curve follows from that of $H(\theta)$. Where H is largest
the steel is saturated and so μ is small. Conversely where H is smallest
μ does not reach its peak value and so is again small. For intermediate values of H, $\mu(\theta)$ moves back and forth across the maximum of μ.

A similar procedure was carried out for the cases $R_2 = 3.5$ and $R_2 = 4.0$.
A useful measure of the performance of the screen is the 'Attenuation
Factor' 'A' defined as:

$$A = (|H| \text{ unscreened})/ (|H| \text{ screened}).$$

Figure 7 plots this parameter for the three cases considered above.

Finally, the whole procedure was repeated using the $\mu(H)$ curve for
Data 2 in Fig 2. Fig 8 gives the resulting attenuation in this case.

Optimisation of the spherical screen

Figs 7 and 8 show that the best overall screening is obtained using a
screen of radius 3.5m. This is true for screens of permeability given
by either Data 1 or Data 2 (see Fig 2). Thus there is a basic

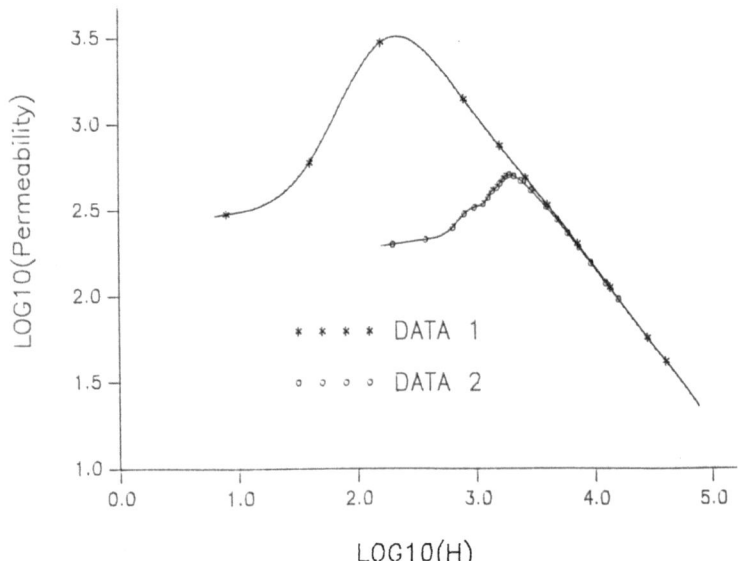

FIGURE 2. CUBIC SPLINE FITS TO PERMEABILITY DATA FOR MILD STEEL AND
A HIGH ALLOY STEEL

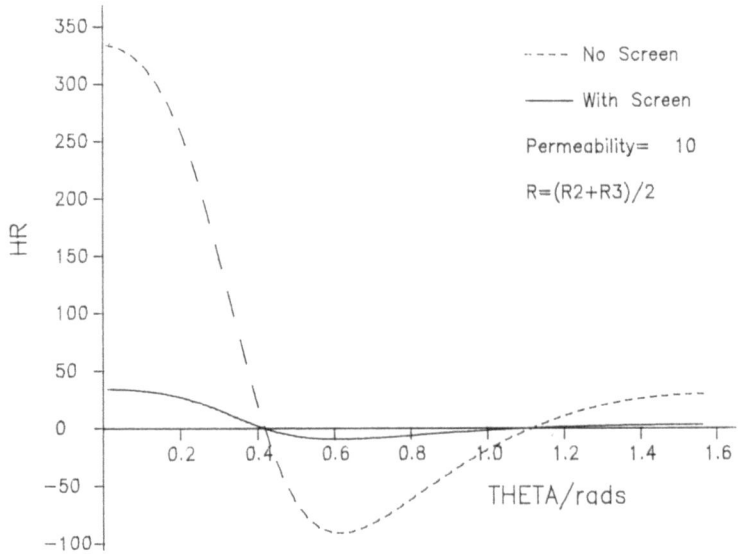

FIGURE 3. VARIATION WITH ANGLE OF RADIAL FIELD IN SCREEN

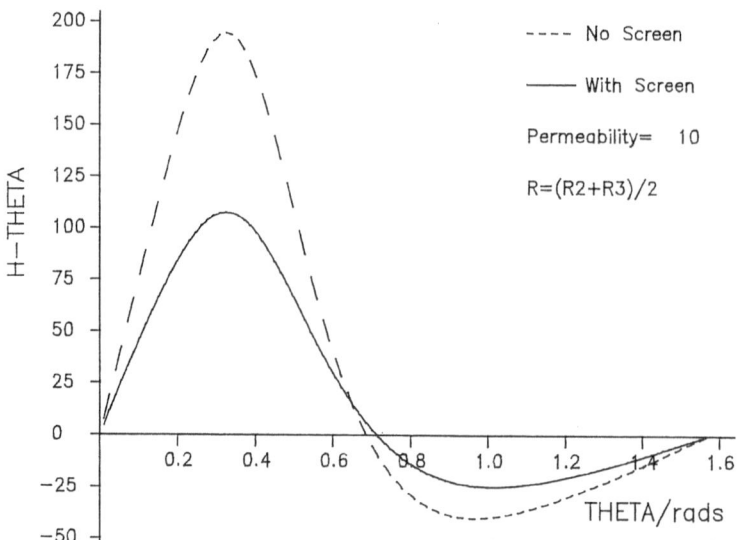

FIGURE 4. VARIATION WITH ANGLE OF AZIMUTHAL FIELD IN SCREEN

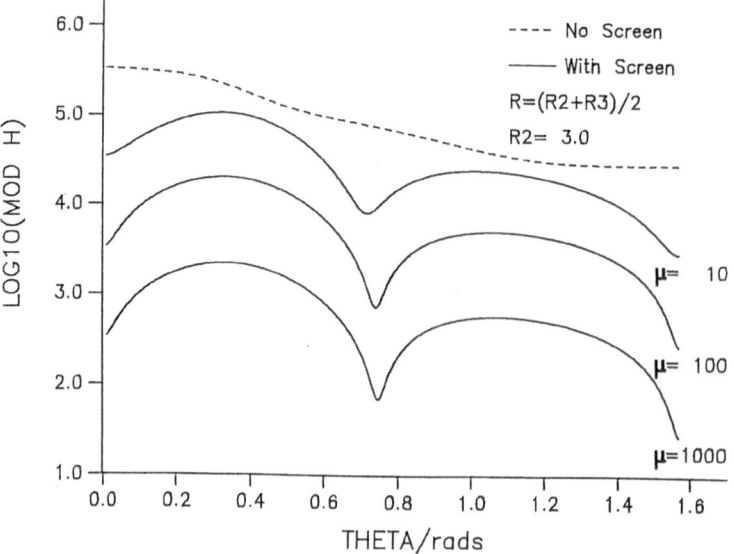

FIGURE 5. FIELD IN SCREEN AS A FUNCTION OF SCREEN PERMEABILITY

162

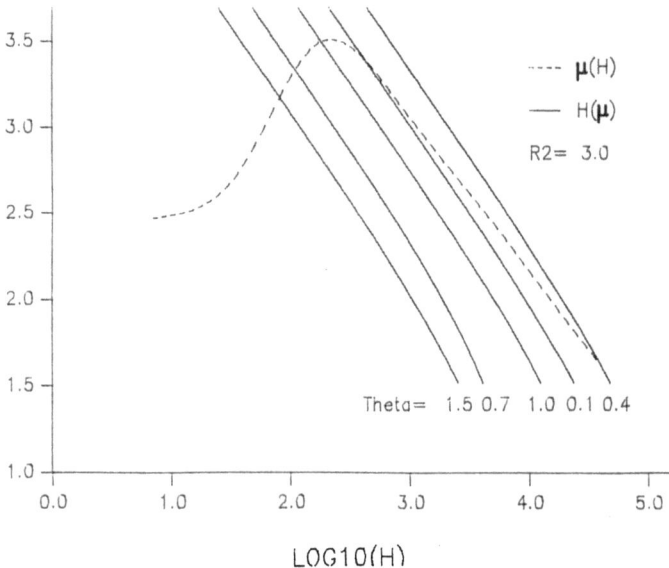

FIGURE 6. CONSTRUCTION FOR CALCULATING THE ACTUAL BEHAVIOUR OF THE SCREEN

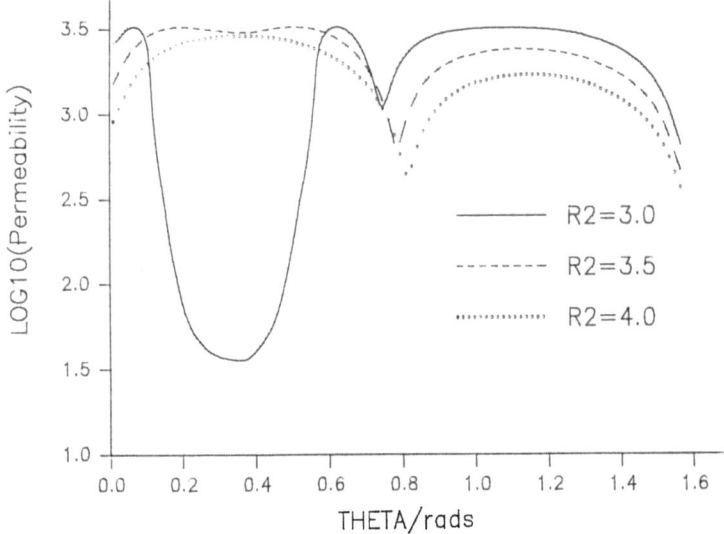

FIGURE 7. EFFECT OF SCREEN IN ATTENUATING STRAY FIELD – SCREEN PERMEABILITY GIVEN BY DATA 1

similarity of performance of the two screens, in spite of the large difference in their permeability behaviour. A more detailed examination of the results shows that the optimum screen configuration occur when the screen material has its peak permeability in the region of t screen where the magnetic field is a maximum. Thus, the existence of an optimum screen radius is associated with a maximum in the permeabi ity vs. field curve for the screen material.

4. DISCUSSION

The results shown in the previous section have indicated that an opt-imum radius exists for the location of a magnetic shield. The exist-ence of this optimum is related to the existence of a maximum in the permeability of the shield material. Although most of the discussion in this paper has involved the use of a material with a particular B-characteristic a maximum in the permeability is exhibited by many fer -magnetic materials. Hence the existence of an optimum shield is a result of general applicability.

The method used in this paper seems to suffer from a basic inconsist-ency in that the initial equations assume a shield of constant perme-ability μ. However the results can only be made consistent with the measured μ v. H characteristic of the shield material by assuming tha μ varies with angle. An inspection of Fig 7 shows that, for the opt imum shield arrangement, the variation of μ with angle is small for t greater part of the shield. The scalar potential V can only be expre -ed as a Legendre polynomial expansion if it satisfies Laplace's equation. Within the iron shell, however,

$$\nabla^2 V = \underline{H} \cdot \nabla\mu \quad \text{(as explained in Section 2)}$$

$$= H_\theta \frac{d\mu}{d\theta}$$

since μ is a function of θ only. On comparing Figs 4 and 7 it is evident that (excluding $R_2 = 3.0$) H_θ becomes zero at the angle for which $\frac{d\mu}{d\theta}$ is significant. Thus the deviation from Laplace's equation is small and the method used in this paper is justified a posteriori

5. SUMMARY

In this paper, a simple method has been presented for calculating the effect of a ferromagnetic screen in reducing the stray magnetic field due to a field coil system. This method has been applied to a typica case, and the effect of altering the screen has been examined. Using practical materials, it appears that an optimum screen configuration exists.

References

1. Culwick B B, Brookhaven National Lab. Rep. No. HHO5-O,
   BBC-1, 1965 (unpublished).

2. Garrett M W, "High Magnetic Fields", H Kolm et al., Eds.
   New York: Wiley, pp. 14 - 26, 1962.

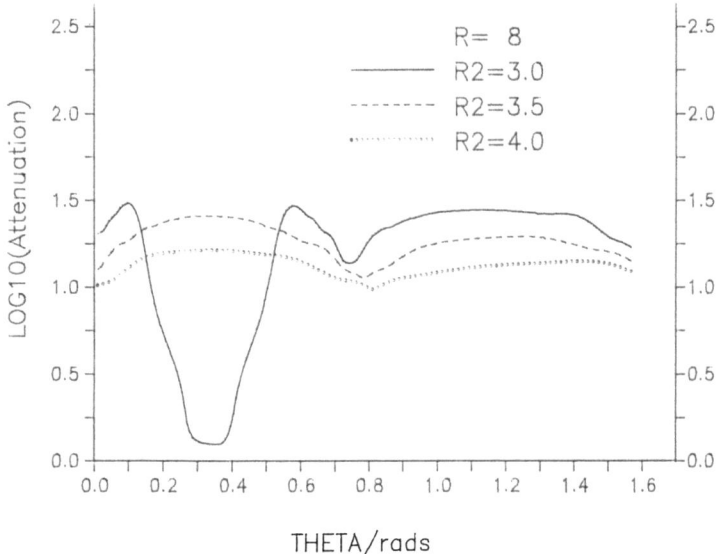

FIGURE 8. EFFECT OF SCREEN IN ATTENUATING STRAY FIELD - SCREEN
PERMEABILITY GIVEN BY DATA 2.

PERFORMANCE CALCULATIONS FOR DEVICES WITH PERMANENT MAGNETS

D Howe and W F Low

University of Sheffield

1. INTRODUCTION

The development of new and improved grades of permanent magnet materials has led to a major reassessment of their use and of the design methods employed. They are now being used increasingly in applications as diverse as electrical machines, actuator systems, microwave and charged particle beam devices, and the medical sciences. Although rare-earth cobalt is replacing older established materials in many existing devices, and allowing applications which were previously not feasible, it currently accounts for only about 10% of the magnet market, of which about 50% is still held by the relatively low cost ferrites, and the remainder by metal magnets largely of the Alnico type.

The new and frequently more exacting uses of magnets has made the traditional design approach, which was based on relatively crude lumped circuit calculations, inappropriate. More accurate techniques, based on numerical methods such as finite elements, are often essential to account for the various material characteristics on which magnets may be operated. It is the range and shape of these characteristics which makes magnet design so varied and demanding, and which requires a design method if it is to be of general application, with the flexibility to handle the many aspects of magnet working.

In order to illustrate the different conditions under which magnets are commonly operated the paper considers some practical permanent magnet systems, which are analysed by a suite of finite element programs capable of accounting for the magnetic and thermal history of magnets.

2. TYPICAL OPERATING CONDITIONS

Figure 1 shows part of a typical hysterises loop for a permanent magnet. A general rule is that the minimum field required to ensure complete magnetisation is 3-5 times the coercivity H_c. On removal of the magnet from the magnetising field it will be demagnetised to a point C on the major curve, which depends on the demagnetising force to which it is exposed. If, subsequently, it is returned to a device the demagnetisation curve is not retraced, but follows a recoil loop, which since it is narrow may be considered as a straight recoil line CD of constant slope equal to the recoil permeability. The final working point D will depend upon the configuration of the device for which the magnet provides the excitation.

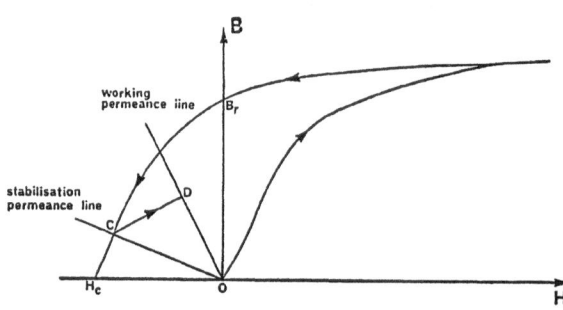

Figure 1. <u>Magnetisation, stabilisation and recoil working of permanent magnet</u>

In general therefore a permanent magnet involves three stages in the calculation of its performance; the initial magnetisation stage, to provide a localised information on the direction and level of magnetisation attained; the stabilisation stage, to determine the limiting operating points on the demagnetisation characteristic when under the influence of demagnetising fields; and the recoil stage, to establish the operating points following stabilisation. The need to perform all three stages will depend upon the particular type and geometry of magnet and its application.

Types of permanent magnet

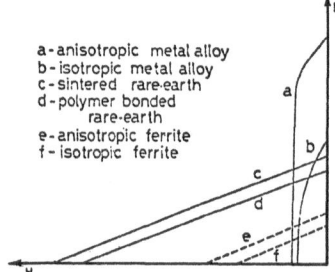

a - anisotropic metal alloy
b - isotropic metal alloy
c - sintered rare-earth
d - polymer bonded
    rare-earth
e - anisotropic ferrite
f - isotropic ferrite

Figure 2. <u>Magnet characteristics</u>

Figure 2 shows the commonly available major demagnetisation characteristics of the principal magnet materials. Since the ceramic ferrites and the rare-earths have an essentially linear characteristic they can often withstand demagnetising fields without any irreversible loss of flux. Because of this inherent stability it is usually unnecessary to calculate their stabilisation. The metal magnets have a highly non-linear characteristic, so that after exposure to a demagnetising field they will work under recoil conditions having been irreversibly demagnetised, though stabilised against further demagnetisation by lower fields. The anisotropic materials, which offer improved performance both in terms of remanence and coercivity, are characterised further by their magnetisation characteristic in the transverse direction. Figure 3, for example, shows magnetisation curves in the preferred and transverse directions for anisotropic barium ferrite. Under static conditions the magnet would work on its major curves, whilst under dynamic operation it would work under recoil conditions. Unless the magnet experiences a significant component of flux in the transverse direction, however, the permeability assigned to that direction will not be critical in calculations of the field (Binns, 1975).

168

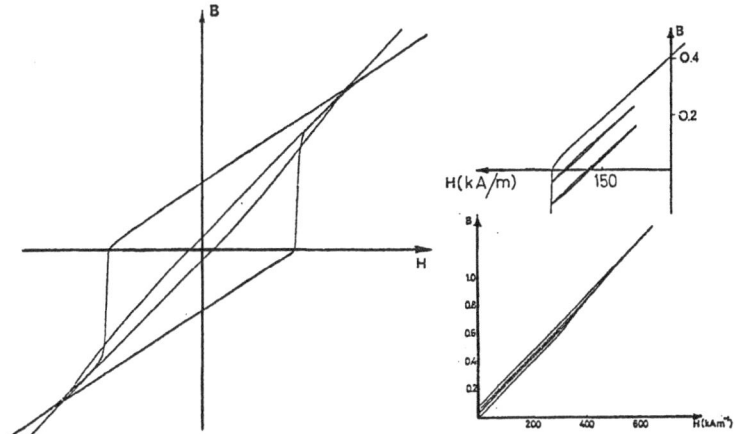

Figure 3. Magnetisation curves for barium ferrite in preferred and transverse direction

Incomplete magnetisation

For ease of production, magnets are often magnetised in situ and consequently the applied field may not be such as to fully magnetise all parts of a magnet in the desired direction. Further, to simplify device construction magnets may be cast into complex shapes and again full magnetisation is unlikely. Hence it is often important to incorporate minor curves in the calculation of the field under subsequent stabilisation. A number of such curves are shown in Figure 4 for anisotropic barium ferrite.

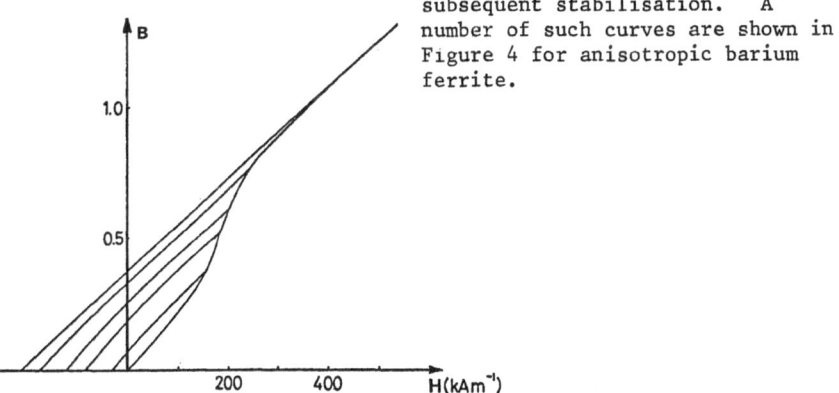

Figure 4. Minor loops for anisotropic barium ferrite in preferred direction

Demagnetising fields

The demagnetising effect of an external field in the direction of initial polarisation, due to the curvature of the demagnetisation curve in the second quadrant, has been illustrated by Figure 1. More severe external fields may drive the working point into the third quadrant. Hence to model the demagnetising effect it would be necessary to include third quadrant working, even though the magnet may have a linear characteristic in the second quadrant. Further fo magnets that experience demagnetisation due to armature reaction in electrical machines it would also be necessary to include first quadrant working as shown in Figure 5 to account for the magnetising component of the armature reaction field.

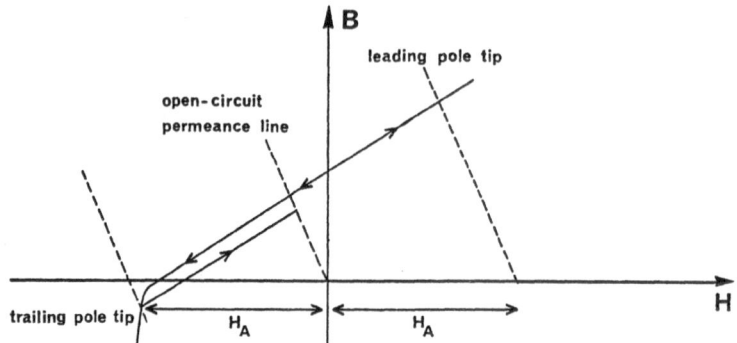

Figure 5. Effect of an armature reaction field in a D.C. motor

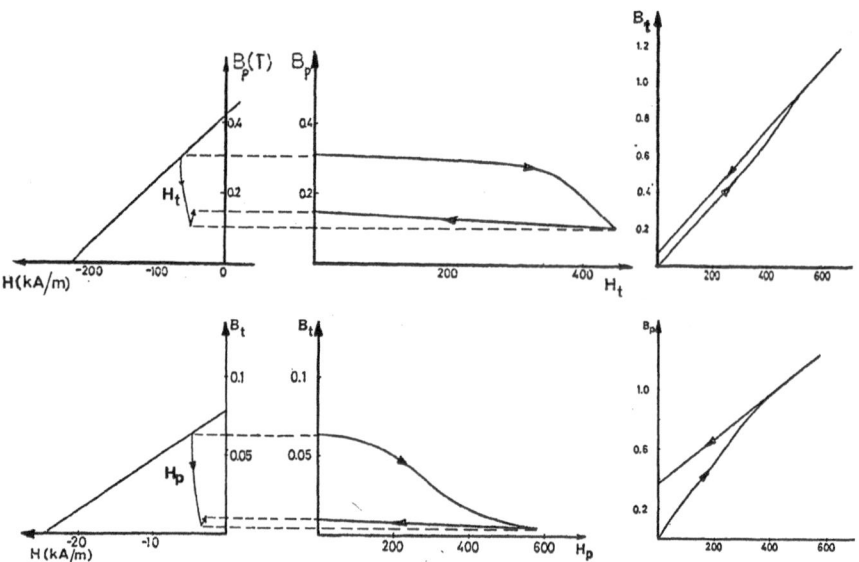

Figure 6. Behaviour of anisotropic barium ferrite under excitation in two directions at right angles

If a magnet is exposed to an external field which causes the field within the magnet to deviate substantially from the preferred direction of magnetisation the component of field in the transverse direction may have a significant irreversible effect on that in the preferred direction, and vice-versa.   Figure 6 illustrates this behaviour for anisotropic barium ferrite.   The measurements show that, after initial magnetisation along either the preferred or the transverse directions, such a magnet can be irreversibly demagnetised in the direction of initial magnetisation and permanently magnetised in the direction in which the external field is applied.   The magnet is then stabilised against further demagnetisation by lower external fields.   Again such information is relevant to magnets which directly experience armature reaction fields.

Temperature effects
The characteristics of all permanent magnets vary with temperature, particularly those of ferrites.   This variation may be reversible or irreversible depending on the temperature and the working permeance.   Above the temperature in the reversible zone the properties are permanently impaired and become completely destroyed at the Curie temperature. Figure 7 shows the effect of temperature on the characteristic of orientated strontium ferrite (Kippax, 1967).   When the working permeance line intersects the straight portion of all the characteristics the effect of thermal cycling is reversible.   When the line intersects some of the characteristics beyond their linear range an irreversible

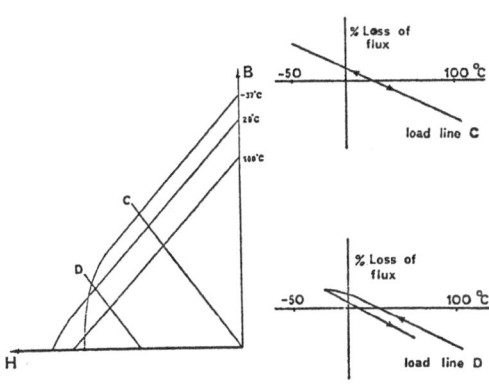

Figure 7. Effect of thermal cycling

change occurs, although subsequent cycling within the same temperature limits will give a reversible change.

Other considerations
Although permanent magnets are structurally stable when operated within specification and largely insensitive to mechanical stress, sintered magnets can exhibit stress anisotropy after pressing which will influence their magnetic properties.

Of course, all the conditions outlined above will not normally need to be considered when predicting the performance of many permanent

171

magnet devices. For example, since rare-eath magnets are manufact-
ured in simple shapes which, due to their inherent stability, are
usually fully magnetised before assembly, it is necessary to consider
only operation on their major characteristic. Further, since they
will have virtually uniform magnetisation they can often be modelled
by an equivalent surface density distribution so that analytical solutions for the resultant field distribution are sometimes possible.
However most practical devices will be too complex to permit exact
analysis and numerical methods must be used, of which the most convenient is the finite element technique.

3. FINITE ELEMENT ANALYSIS

The authors have developed a suite of finite element programs which
provide the option for computing the field of the various types of
permanent magnet under one or more of the operating conditions described earlier. They make it possible to account for the magnetic
and thermal history of a magnet since information characterising each
element at any stage is transferred by disc file between the various
programs. The programs can handle two-dimensional, planar and axisymmetric, as well as three-dimensional problems. The resulting non-
linear algebraic equations are solved by Newton-Raphson iteration.
For two-dimensional problems the linear equations formed on each
iteration are solved by a banded Gaussian Elimination technique, whilst for three-dimensional problems a pre-conditioned conjugate gradient method, in which pre-conditioning is achieved by scaling the coefficient matrix symmetrically, is used.

Space does not permit anything other than a brief outline of the
modelling techniques. Basically the approach adopted is to represent the field in the permanent magnet by

$$\begin{vmatrix} B_p \\ B_t \end{vmatrix} = \begin{vmatrix} a & b \\ c & d \end{vmatrix} \begin{vmatrix} H_p \\ H_t \end{vmatrix} + \mu_o \begin{vmatrix} M(B_p) \\ o \end{vmatrix} \tag{1}$$

where the subscripts p and t refer to the preferred direction of
magnetisation and the transverse direction respectively. It is
through the choice of the elements a, b, c, d and $M(B_p)$ that different aspects of magnet working can be represented.

Equation (1) can be expressed in terms of a cartesian coordinate system in two-dimensions by

$$\begin{vmatrix} B_x \\ B_y \end{vmatrix} = \begin{vmatrix} A & B \\ C & D \end{vmatrix} \begin{vmatrix} H_x \\ H_y \end{vmatrix} + \mu_o \begin{vmatrix} M_x \\ M_y \end{vmatrix} \tag{2}$$

where A $= a \sin^2\alpha + d \cos^2\alpha - (b+c)\sin\alpha\cos\alpha$

B $= (a-d)\sin\alpha\cos\alpha + b \sin^2\alpha - c \cos^2\alpha$

C $= (a-d)\sin\alpha\cos\alpha - b \cos^2\alpha + c \sin^2\alpha$

D $= a \cos^2\alpha + d \sin^2\alpha + (b+c)\sin\alpha\cos\alpha$

$$M_x = M(B_p)\sin\alpha$$
$$M_y = M(B_p)\cos\alpha$$
$$\alpha = \text{angle of magnetisation wrt y-axis}$$

The corresponding functional when the field components are derived from a single component vector potential function A_z is

$$F = \int_R \left\{ \left[\left(\frac{A'\ \partial A_z}{\partial x} + \frac{C'\ \partial A_z}{\partial y} \right) \frac{\partial A_z}{\partial y} + \left(B' \frac{\partial \phi}{\partial x} + D' \frac{\partial \phi}{\partial y} \right) \frac{\partial \phi}{\partial y} \right] \right.$$
$$\left. - 2JA_z + J_m \right\} dR \qquad (3)$$

where $J_m = 2\mu_o \left[(A'M_y - C'M_x) \frac{\partial A_z}{\partial x} + (-D'M_x + B'M_y) \frac{\partial A_z}{\partial y} \right]$

$$A' = A/(AD-BC)$$
$$B' = B/(AD-BC) \quad \text{etc.}$$

A similar approach can be followed for a scalar potential formulation.

In equation (1) the element a is equal to μ_o, element d depends on whether the operating condition in the transverse direction is static or dynamic, and the off-diagonal elements b and c will be zero if B_p and B_t are mutually independant.   When the magnet works on the major demagnetisation characteristic, the function $M(B_p)$ takes the form

$$\frac{B_p}{\mu_o} - \frac{H_c(B_r-B_p)}{B_r(1+\beta B_p + \gamma B_p^2)} \qquad (4)$$

where
$$\beta = \frac{H_c - 2H_m}{B_m H_m}$$

$$\gamma = \frac{H_m B_r - B_m H_c}{B_m^2 H_m B_r}$$

and B_m, H_m = B and H at BH_{max}

When working under recoil conditions $M(B_p)$ becomes

$$\frac{B_p}{\mu_o} - \frac{B_p - B_{stab}}{\mu_o \mu_{recoil}} + H_{stab} \qquad (5)$$

where B_{stab}, H_{stab} = B and H at the stabilised operating point
μ_{recoil} = recoil permeability.

In order to account for the interaction of the components of the field in magnets under excitation in two directions at right angles, elements b and c could be assigned values to represent the observed behaviour (Binns, 1983). Alternatively $M(B_p)$ can be modified during the course of the iterative solution.

Of course, although field calculations are the key to design procedures it is essential to compute global parameters from which the performance can be predicted. The programs therefore include the facility for calculating forces and torques through integration of the Maxwell Stress Tensor

$$\underline{T} = \mu_o \left[(\underline{n}.\underline{H}) \underline{H} - \frac{H^2}{2} \underline{n} \right] \tag{6}$$

over defined surfaces, and winding inductances under operating conditions determined by the permeability distribution selected.

4. DESIGN EXAMPLES

The examples which follow will serve to illustrate the application of the f.e. programs to account for some of the different aspects of magnet working which influence the performance of practical devices.

Electrical machines
An important factor in d.c. motors is the resistance of the magnets to demagnetisation by armature reaction mmfs, the significance of which is heavily dependant on machine size since the maximum armature AT/pole tends to increase as (pole-pitch)^2. Figure 8 shows the field distribution in a 2-pole motor, rated at 1kW FL and equipped with radially orientated barium ferrite segments, before, during and after a high armature current input condition. It will be seen that the magnets can withstand a certain level of demagnetising armature reaction field before the working point of elements on the trailing pole tip is driven beyond the linear part of their demagnetisation curve, and that demagnetisation is more severe at the lower temperature. Figure 8 also shows the field distribution due to armature currents alone, assuming the permeability distribution in the stator and rotor iron to be unchanged from that on o/c and the magnets to have a constant recoil permeability, from which the slot component of armature winding inductance can be calculated. The suite of programs has allowed the effects of design changes, such as the use of composite magnet segments and a 'split' stator core, aimed at raising the demagnetisation withstand to be evaluated (Howe, 1983).

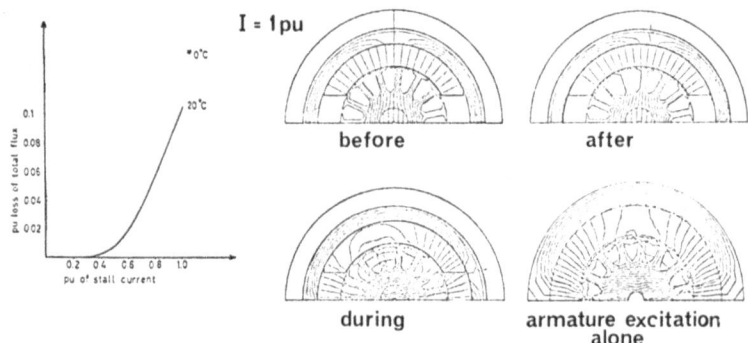

Figure 8. Permanent magnet d.c. motor

Figure 9 shows part of the cross-section of a 6-pole, 1-ph generator
with excitation from pole-face mounted rare-earth magnets on the rotor.
Because the effective permeability of permanent magnets is low the
equivalent circuit simply comprises the excitation emf $k\omega$ behind the
winding inductance L and resistance R.   The generator supplies a

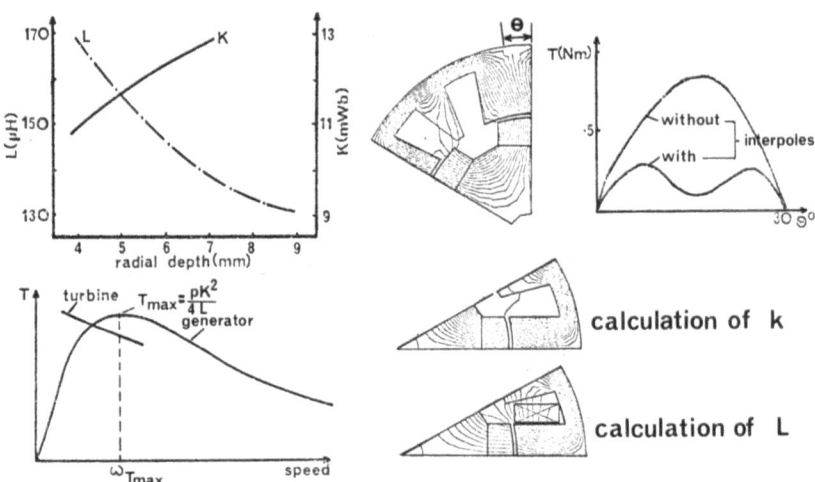

Figure 9.   Permanent magnet a.c. generator

tungsten lamp R_L and is driven by an air-turbine.   In order to match
the generator to the turbine to give the required output at a specified speed and air-pressure it is necessary to select a suitable magnet geometry which gives the required stator winding flux linkages k

175

and the correct combination of k and L by using the curves in Figure which show the effect of varying the dimensions of the magnets (Tan, 1981). Figure 9 also highlights the effectiveness of stator inter-poles in limiting peak saliency torque.

Dynamic Sensors

The displacement which results when an input is applied to a gyro-scope or accelerometer is sensed by a pick-off whose output results a current being passed through a torquer coil to interact with the field in a permanent magnet circuit to reduce the displacement to zero.

Figure 10 shows a single-axis linear accelerometer equipped with a 'voice coil' torquer in which the airgap field is produced by a diametrically magnetised non-columnar Hycomax III magnet. In order to allow for 'end effects' it is necessary to determine the field distribution in 3-d. It will be seen that there is some advantage in allowing the magnet to overhang the yoke.

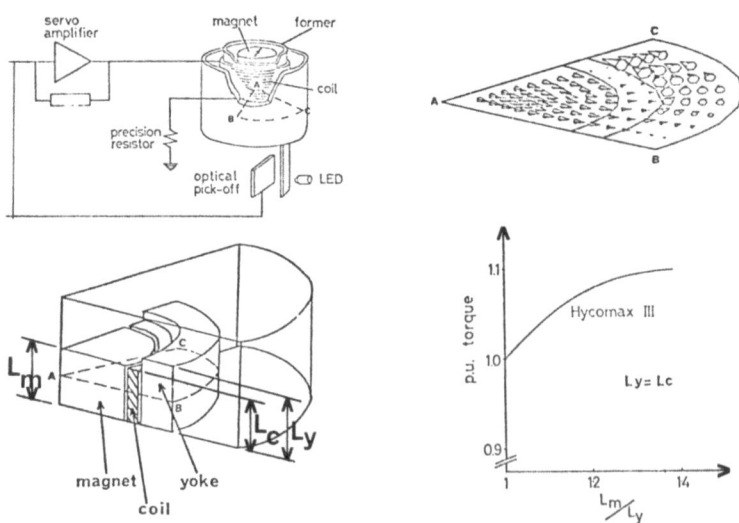

Figure 10. Servoed linear accelerometer

Figure 11 shows a gyroscope having sensing capability about 2 axes. The torquer configuration is based on two annular samarium cobalt magnets so that both sides of each torquer coil are effective in producing torque. However, with this arrangement a torque can be developed when the magnets are displaced relative to the ferromagne-tic gyro case even when the torquer coils carry no current, except

176

for one particular magnet separation.   In order to calculate this separation, it is again necessary to calculate the magnetic field distribution in 3-d, as shown in Figure 11, (Low, 1983).

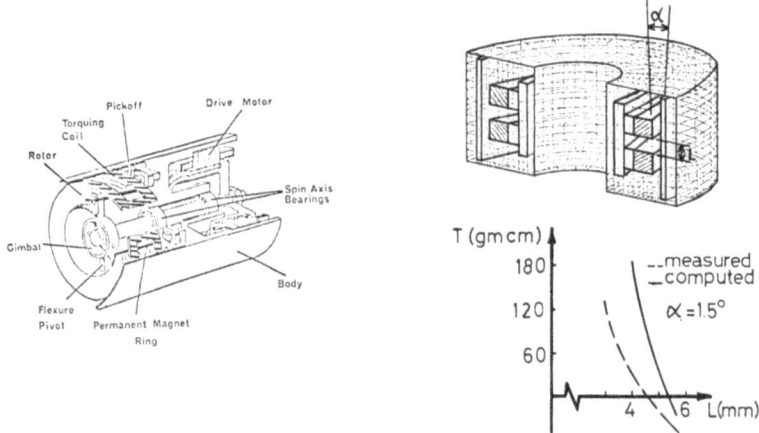

Figure 11.   Two-axis gyroscope

Microwave Devices

Figure 12 shows examples of recently designed magnet systems for magnetrons and kylstrons, where miniaturisation has been achieved through the use of rare-earth magnets (Howe, 1983).   The devices shown all possess rotational symmetry.   In each configuration the

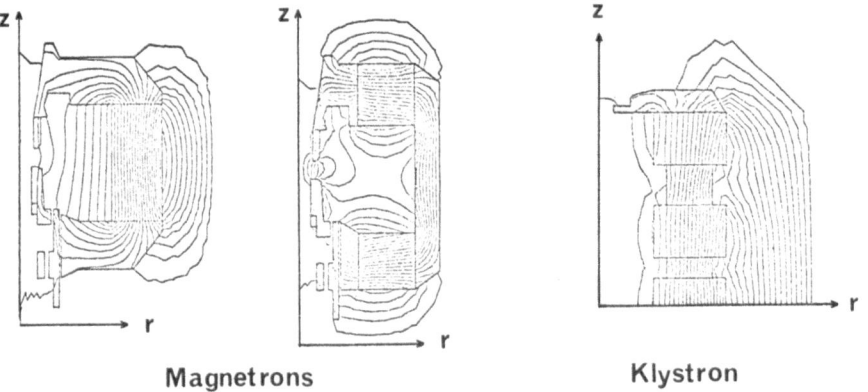

Figure 12.   Microwave devices

aim is to work the magnets at around their optimum BH_{max} point, and to achieve the desired level and uniformity of flux density in the interaction space by a focusing system.

5. CONCLUSIONS

The paper has shown that the design of parmanent magnet devices requ
careful consideration of many facets of the effects of their charac
istics. However, by employing a package such as that described it
possible to account for the magnetic and thermal history of the magnets in performance calculations. The programs could be used as a
routine design aid or to develop generalised design rules to simplif
the selection of parameters for a particular device.

6. REFERENCES

Binns, K.J. Jabbar, M.A. Barnard, W.R: Computation of the magnetic
field of permanent magnets in iron cores, Proc. IEE, 122(12), 1377-
1381, 1975.

Kippax, J.I: The effect of temperature on the magnetic properties
of orientated strontium ferrite, Mullard Magnet Applications Lab.
Report 9, 1967.

Howe, D, Birch, T.S: Armature reaction in permanent magnet d.c.
motors with ferrite magnets, Proc. 18th UPEC, 697-703, 1983.

Tan, G.H, Howe, D, Birch, T.S: Design and performance prediction
of a permanent magnet air-turbo generator, IEE Conf. Pub. 202,
125-128, 1981.

Low, W.F, Howe, D, Birch, T.S, Capaldi, N.R: Dynamic sensors for aer
space applications, Proc. 18th UPEC, 705-711, 1983.

Howe, D, Birch, T.S: Effective replacement of metallic magnets by
rare-earth cobalt, Proc. 6th Int. Workshop on RE Cobalt, 21-30, 1982.

Binns, K.J. Low, T.S, Jabbar, M.A: Behaviour of polymer-bonded
rare-earth magnet under excitation in two directions at right angles,
Proc. IEE, 130(B), 1, 25-32, 1983.

THE USE OF COMPLEX PERMEABILITY FOR STEADY STATE NON-LINEAR EDDY-CURRENT AND HYSTERESIS PROBLEMS

A G Jack[1] M R Harris[1] and P T Jowett[2]

1 University of Newcastle upon Tyne
2 Ferranti Ltd., Edinburgh

1. INTRODUCTION

There are many problems in electromagnetics which involve a steady state field in a ferromagnetic material. Often the material is magnetically soft and the hysteresis of the material is ignored. Further simplifications involving the choice of a single permeability are made so that the problem can be reduced to the solution of single harmonic.

A full solution of such problems requires either a multiple harmonic solution or, more usually, a time stepping solution involving a full representation of the B-H loops for the material. This is both expensive and complicated. A cheaper approach is to allow only a single harmonic response to excitation in the same harmonic. A complex permeability which is a function of peak magnetic field strength can then be defined which takes account of hysteresis and magnetic non-linearity.

2. BASIC METHOD AND RANGE OF APPLICATION

The complex permeability method is applicable to computational problems that are cyclic in time, the essence of the approach being to neglect the time harmonics of B and H fields that are produced in general at all points in space. The phasor relationship between the fundamental components B_1 and H_1 is then expressed by the complex permeability, $\mu_r \mu_o \exp(-j\phi)$. Both μ_r and ϕ (the hysteretic lag-angle of B_1 with respect to H_1) are functions of the peak value of local magnetic field, H, and are readily determined by Fourier analysis of a family of quasi-stationary B-H loops for the material in question. That is, a sinusoidal H is assumed and from this the resulting variation of B is determined from the relevant B-H loop. This is Fourier analysed and the complex ratio of H fundamental to B fundamental gives the complex permeability. Clearly a sinusoidal B could be assumed and H fundamental obtained by Fourier analysis. This would be more relevant for voltage forced solutions.

Because of the neglect of harmonic components, however, it is unavoidable that the local values of μ_r and ϕ are selected, not according to the

179

peak of the total H-field, but the peak of the fundamental component only, H_1. This shortcoming of principle is one source of error; another, in part tied up with the first, is the neglect of power losses that would arise directly from the presence of the harmonic fields. Specifically, these include harmonic eddy-current loss, and hysteresis loss in minor loops (if saddles occur in the H-waveform).

Despite these sources of error, and despite difficulty in the treatme of magnetic viscosity and rotational hysteresis (in cases where these effects occur - see later), it is the author's view that the method h much in its favour. In a study of a one-dimensional non-linear and strongly hysteretic diffusion problem, briefly reported in this paper accuracy of a few per cent in the estimation of power losses has been generally achieved. There are grounds for thinking that similar accuracy might be obtained in application to finite element analyses in two and three dimensions, where it would appear that little if any use of the method has so far been made. Yet the potential saving in computational effort is broadly about an order of magnitude, as compa with the fully non-linear time-stepping solution for the same case. is therefore hoped to encourage other workers to examine the possible usefulness of this approach, as indeed the present authors will be do: in a wider variety of diffusion field problems.

The method has been extensively examined by O'Kelly (1972), specifica for the one-dimensional diffusion problem with surface excitation by uniform sinusoidally varying H-field, the alternative of a travelling surface field being considered in later work. His conclusions for th accuracy of the method in respect of loss, by comparison with experiments, are the same as in this paper. However, detailed loss results have only been published for mild steel, which has a narrow hysteresis loop and generally exhibits the property: hysteresis loss/ eddy-current loss <<1, for any problem with pronounced skin effect. method has been used for high-hysteresis material in the analysis of machines (O'Kelly 1976) with apparently satisfactory accuracy, but de of the loss calculation is not reported. The particular material (35 cobalt steel) also exhibits magnetic viscosity, the dimensions of the hysteresis loops changing with frequency over the low-frequency range and correction for this effect must to some extent cloud the issue of the inherent accuracy of the complex permeability approach.

Demerdash and Gillott (1974) propose and use an approximate permeabil function in magnetic network studies of machines, derived by equating average stored energy densities in exact and approximate representati They do not observe - but it can in fact be readily shown - that thei definition is identical to that described in this paper, but with $\phi =$ Their work thus introduces the important principle of a harmonic permeability applied in a non-layer type of problem (as distinct from O'Kelly), though not in a full-blown finite element analysis, and wit hysteresis constrained to be zero. They also report good accuracy in

180

their results, for fundamental flux density distribution, and for active and reactive power.

The principle of discarding harmonics and concentrating on the fundamental, is quite reminiscent of the describing function technique, widely used in non-linear control theory (Atherton 1975). That technique enjoys success in many practical problems, though always (as here) one must be satisfied that the harmonics are not of dominant importance.

3. THE PROBLEM STUDIED

In this paper, the authors have taken the case of 3% cobalt steel, which gives roughly 18 x more hysteresis loss than mild steel for the same degree of skin effect, and has been found experimentally not to exhibit viscosity. The specimen is in the form of a ring, toroidally wound, of outer diameter 19.18 mm, radial depth 1.28 mm, and axial length 12.27 mm. Experimental results for loss are available, and show encouraging agreement with calculations, but the paper concentrates on a comparative evaluation of two methods of computation, namely the complex permeability method and an 'exact' non-linear hysteretic time-stepping solution.

The specimen is considered in linearly developed form in both computations, with axial end-effects neglected. There is no provision in the time-stepping program for flux excursion on minor hysteresis loops, but the evidence suggests that this is an insignificant restriction; algorithms for its inclusion have been developed. Difficulty over the difference between rotational and pulsating hysteresis losses does not arise in this simple configuration. Computations are taken up to 5 kHz, at which point a rough estimate of the equivalent classical skin depth shows it to be about 23% of the ring depth; as will be seen, about 30-35% of the total loss is still hysteresis. A sinusoidally varying, uniform surface H-field is assumed applied to both faces of the specimen, corresponding to sinusoidal exciting current. (This matches the assumption that is made in deriving the basic μ_r, ϕ data; as noted earlier it would be possible to derive alternative data, for example on the basis of sinusoidal B, and the practical significance of this choice will be explored in future work).

Fig 1 shows the family of experimental B-H loops for the material, from which the complex permeability characteristics of Fig 2 are simply obtained by Fourier analysis. Conductivity, $\sigma = 37.0 \times 10^5$ S/m at 25°C.

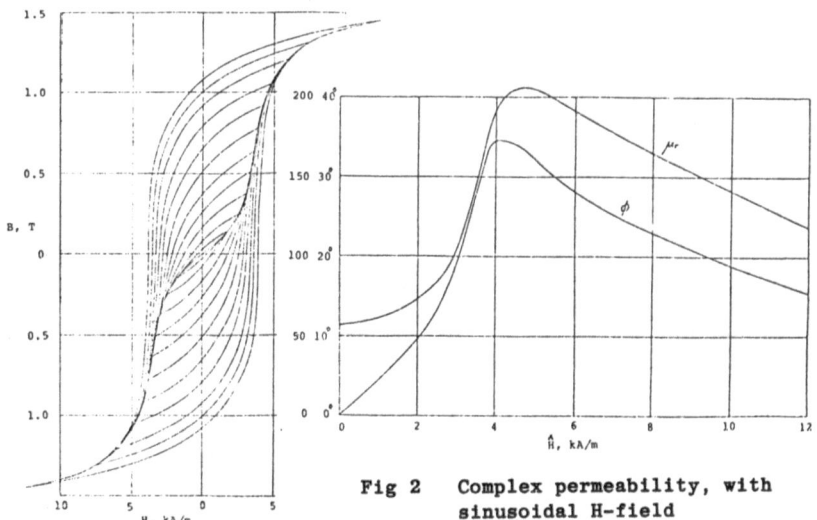

**Fig 2   Complex permeability, with
sinusoidal H-field**

**Fig 1   Hysteresis loops for
3% cobalt steel**

Time-stepping is by a centred-difference, Dufort Frankel, explicit
method, based on the equation

$$\text{curl curl } H = -\sigma \; \frac{\partial B}{\partial H} \; \frac{\partial H}{\partial t}$$

For this purpose, B-H data is stored in the form $\partial B/\partial H$ vs H, using
cubic splines with linear interpolation; bi-cubic splines will be use
in later work. The choice of an explicit method was influenced by
earlier work (Jack et al 1978) on non-linear eddy current problems
without hysteresis. The formulation entirely in H was chosen in orde
that a centred difference explicit method could easily be used. The
choice of formulation strongly influences the severity of the numeric
solution algorithm.

Complex-permeability solutions are easily obtained for this particula

182

problem, without iteration, by assuming values of H_1 at the centre of
the specimen and computing through the layers to the face, to discover
the corresponding surface-H for each case. The program was verified
against the solutions given in O'Kelly 1972.

4. RESULTS

Results for total ring loss by complex permeability are shown as the
plotted curves in Fig 3, at four different frequencies. The circles
represent five spot calculations by time-stepping and the crosses
measured results. The highest frequency is displayed on Fig 4, with
hysteresis and eddy-current components shown separately, in addition
to the total.

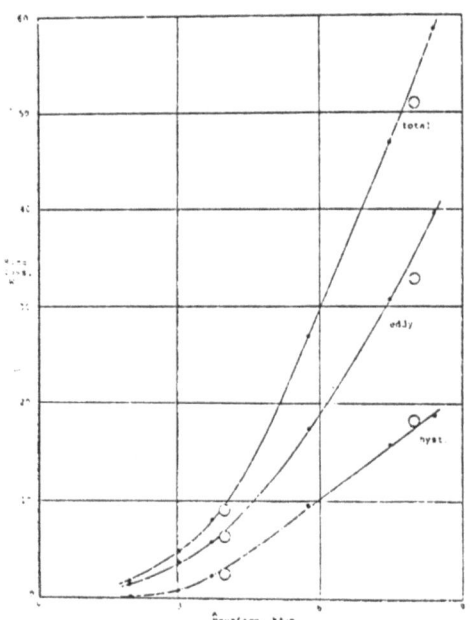

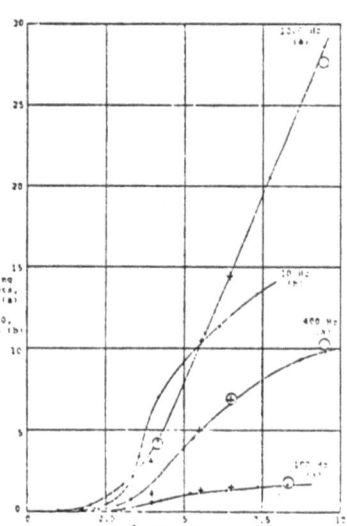

Fig 3   Loss characteristics
up to 1 kHz

Fig 4   Eddy current and hysteresis losses
at 5 kHz

183

Table 1 displays hysteresis and eddy-current data at all frequencies, by both methods. In the time-stepping case, for the sake of interest total loss is calculated by surface Poynting vector and the component obtained by summing volume integrals; there is consequently a small difference between the sum of the components and the total.

It will be seen that agreement between the two methods is uniformly good for all results. The complex permeability approach appears to work well, and it deserves careful evaluation in application to problems of a more general nature.

f,	H_s,	Ring loss, W					
		Complex permeability			Time-stepping		
Hz	kA/m	Eddy	Hysteresis	Total	Eddy	Hysteresis	Total
50	8.16	.062	.725	.787	.137	.682	.825
100	8.33	.24	1.44	1.68	.44	1.37	1.83
400	6.55	2.51	4.17	6.68	2.58	4.28	7.00
400	9.55	3.95	5.89	9.84	4.53	5.65	10.29
1000	4.16	2.01	2.32	4.33	1.79	2.28	4.27
1000	9.46	16.8	11.8	28.6	14.9	11.8	27.7
5000	4.00	6.5	3.0	9.5	6.5	2.6	9.2
5000	8.00	35.0	17.5	52.5	32.7	17.5	50.9

Table 1 Eddy current and hysteresis losses

5. FURTHER APPLICATIONS

The method has been applied to a two-dimensional finite element soluti of eddy-currents induced in a turbogenerator rotor. In this problem the hysteresis loss has been considered unimportant and this makes th permeability real, but a function of H. A standard Newton-Raphson solution method has been used to solve the resulting complex non-line equations.

REFERENCES

O'Kelly, D.: 'Hysteresis and eddy-current losses in steel plates with non-linear magnetisation characteristics', Proc. IEE, 119, (11) pp 1675-76, Nov 1972.

O'Kelly, D.: 'Theory and performance of solid-rotor induction and hysteresis machines', Proc. IEE, 123, (5), pp 421-28, May 1976.

Demerdash, N.A. and Gillott, D.H.: 'A new approach for determination of eddy-current and flux penetration in non-linear ferromagnetic materials', IEEE Trans. on Magnetics, MAG 10, pp 682-85, 1974.

Atherton, D.P.: 'Non-linear control engineering', Van Nostrand, pp 75-164, 1975.

Jack, A.G. and Stoll, R.L.: 'Negative-sequence currents and losses in the solid rotor of a turbogenerator', Proc IEE Vol 127, Part C, No. 2, pp 53-64, March 1980.

SESSION F

IMPACT OF COMPUTER DEVELOPMENT

Chairman

R BRADLEY

Newcastle upon Tyne Polytechnic

HARDWARE-DEPENDENCE OF ELECTROMAGNETICS SOFTWARE

Peter P. Silvester
McGill University, Montreal

SUMMARY

Engineering software is specialised software shaped by user needs, computer hardware, and analytic techniques. Typical systems take 2-3 years to create and remain in use 7-10 years. Since major hardware alterations occur at intervals of 4 years, an analysis system must reside on obsolete hardware for half its lifetime. Transportation to different hardware commonly produces less than ideal results. New analysis methods appear at intervals of 10 years, so software is mathematically only half a generation behind. Current packages are particularly sensitive to questions of man-machine communication, which is very hardware-dependent.

THE SYSTEM LIFE-CYCLE

All engineering software is necessarily highly specialised. Its content is shaped by the needs of potential users as the system designer perceives them, by the analytic techniques available when the system is designed, and by the computer hardware in reasonably common use at the time. All three factors change with time, so that the life-span of any software package is limited.

Good engineering analysis software takes time to design and perfect. A typical large-scale finite element package requires 2-3 years to design, create, and test; it may be expected to remain on the market 7-10 years. Its total life-cycle is therefore about 10-12 years, a relatively long time in a high-technology world.

Hardware Development
Currently, major changes in computer hardware occur about every 3-4 years. This general rule of thumb applies to both processor hardware and to input-output devices. Changes in processor hardware are often unspectacular from an analyst's point of view, since they usually result in established operations being carried out more conveniently, quickly, or cheaply. Changes in input-output devices, on the other hand, tend to have a more revolutionary aspect, since they frequently alter the external appearance of systems fundamentally. For example,

---

the introduction of graphics terminals in the seventies amounted to more than just an increase in plotting speed; it altered the way work was done.

Since the economic life of software systems is substantially longer than the hardware life-cycle, every analysis system must reside on obsolete hardware at least half its lifetime. Conversion to different hardware systems is at best a half-measure, for software structure is very strongly influenced by what the hardware is capable of doing. For example, a light-pen and a graphics tablet have a great deal of superficial similarity in use, but the tablet can be used to digitise drawings and curves directly, the light-pen cannot. Software transported from light-pen environments to graphics tablets therefore ends up merely using the tablet to mimic a light-pen, without taking advantage of its additional abilities.

Alteration of software to make good use of new hardware capabilities often implies not a few alterations, but basic redesign. Ignoring the capabilities of new hardware is thus possible for a few years, but only for a few, because the unused hardware facilities grow to predominate over those actually exploited by the programs. A new start is then indicated.

Analysis Techniques

New numerical analysis methods likely to affect engineering software have appeared at intervals of about 10-12 years. For example, the large systems of simultaneous equations which commonly arise in continuum analysis were solved by relaxation methods in 1955-1965, by sparsity-exploiting Gaussian elimination techniques in 1965-1975, and by preconditioned conjugate gradients since about 1975. Since analytic techniques change in time-spans comparable to those of software systems, software generally lags about half a generation behind.

Alteration of existing programs to employ new analytic methods is much less easy than it looks at first glance, just as hardware changes can be fully exploited only by redesign of software. To pursue the above example, iterative or semi-iterative methods (i.e., relaxation and conjugate gradient methods) are essentially solution improvers rather than solvers; they construct a convergent sequence of iterates beginning from some initial estimate, and can therefore exploit whatever prior knowledge about the solution may be available. In particular, they are very efficient at computing sequences of closely related solutions, using each one as the initial guess for the next. Sparse Gaussian elimination is good at computing a single solution, but it cannot make use of a priori information. Consequently, well-designed software based on Gaussian elimination encourages the user to specify isolated problems, while systems based on conjugate gradient methods should make it easy for him to generate families of successive solutions located on continuous curves in the problem space. Indeed, an ideally designed software system ought to examine the solutions requested by the user, and to sequence them so as to form a continuous curve in problem space.

The choice of analytic technique is strongly dependent on current processor hardware. To continue the above example, Gaussian elimination techniques became popular in the sixties with the advent of good secondary memory (random-access disk drives). The subsequent success of gradient methods owes a good deal to the availability of abundant cheap memory. Indeed one might well surmise that the nineties should see a resurgence of interest in Gaussian elimination, since such techniques ought to reap great benefits from the development of cheap array processors in the eighties.

The System Designer's View

A major factor in determining the characteristics of a CAD system is the system designer's understanding of the user's requirements. It is an axiom of the software design art that the worst way of determining the user's needs is to ask him, for the user's own assessment is too strongly conditioned by the tools available to him in the past. Most engineering analysts are successful technique-appliers, not solution-inventors. Hence users of CAD systems are generally quite good at specifying what fine-tuning alterations to a system might be desirable, but not very good at defining the overall structure or functions of a system.

The relatively long lifetime of software systems in part derives from the large investment involved in system design and construction, but it may perhaps reflect even more the high cost of user training. The issue is not operator training, but the acquisition of enough specialist electromagnetic engineering knowledge to formulate problems and to use the results effectively. Good current software packages take days or weeks of operator training, but months, even years, of user training. Electromagnetic engineering relied heavily on field-theoretic concepts in the late 1800's and early 1900's, then turned to a circuit-theoretic viewpoint in the twenties and thirties. Half a century later, fields-based skills need to be rediscovered and applied to new situations, a difficult and time-consuming process.

COMPUTERS FOR MODELLING AND CAD

The usefulness of a numerical technique to the designer depends on both the ease of implementing it on the available computer hardware, and the form of access the user must have to the hardware in order to use the software effectively. Computing machinery has developed rapidly with improvements in semiconductor technology and a large spectrum of equipment is now available from the microprocessor to the supercomputer.

Appropriateness and Accessibility

For the user, computer systems have two characteristics of critical importance: their appropriateness to the task at hand, and their accessibility when a job needs to be done. These two factors do not necessarily imply similar requirements and do not have equal importance at each stage of a design. The first requires that the algorithms used be matched to the available hardware. In recent years, the both maximum and minimum computer sizes have become more extreme, so there now exists a large range of computing systems balancing availability against power and cost.

The modern supercomputer is derived from earlier computing machines by using each advance in technology to increase computing power whilst keeping system cost approximately constant. These machines are extremely expensive, and it is important to optimise their use of internal processing facilities at the expense of input-output operations. However, such systems can provide very high speed computing with extremely large address spaces. These features can make them attractive for large analysis systems which may solve problems with 100000 variables or more.

The microprocessor is a development in which technological advances have been used to reduce the system cost whilst keeping computing power roughly constant. They are cheap enough that it is no longer important to maximise use. Hence it is possible to dedicate a computer to a single user, thus providing high availability but at the expense of computing speed and with a limited address space.

Modelling software does not depend on hardware directly, but rather on its associated operating system software. But transportable operating systems are technically feasible, which can be operated on two or more different machines. The computing device visible to the user would then appear to be the same in both cases. Of course, the execution speeds of the two are likely to be markedly different! An example of a portable operating system is UNIX, which can be used on some fifty different machines from microprocessors to mainframes.

The consideration of ease of access to the design tools concerns computing hardware and availability of programs for the design engineer. In order for design software to be used on a day-to-day basis it is important to have the computer system readily available to the engineer, thus suggesting at least partial use of low-cost, low-power, micro- or minicomputer based systems in which maximising the utilisation is no longer the most important factor, provided the analysis techniques can be implemented on the system.

Workstations for Interactive Computing

Conventional design of electromagnetic devices is performed interactively; the designer has access to design rules and data, and can refer to these as design progresses. Computer aided design systems should provide the user with either faster access to data previously used, or good access to better data, or, preferably, both.

These two requirements, if applied to the central computing facility, are difficult to meet. The large computer can certainly be used to generate better data and it can maintain a large design data base. However, access is limited and results of any computation are provided either by hardcopy printout and graphs, or on a local terminal. The response to any user request in a large time-sharing environment may well be of the order of minutes rather than seconds. One possible solution to this problem is to provide the design engineer with a local processing system having some disk storage. This machine could provide better response than the timesharing system to the majority of operations, although it may lack the power for handling extremely large data bases.

Ideally, the results of any computation should be presented to the user in a format which he is accustomed to interpreting. Pictures, i.e. graphs, flux plots, loss distributions, etc., can be far more instructive than lists of numbers. Interacting with both the alphanumeric and graphic displays, via some form of pointing device such as a tablet and stylus, allows querying of the results so that only the information relevant to the problem at hand need be extracted. Graphics devices have decreased in cost in much the same way as processors and, as a consequence, high resolution (512 by 512 dots) colour displays are available for even the smallest microprocessors. These can be included as an integral part of an engineering workstation. Thus a typical workstation might consist of a small processor with suitable memory, a raster display device, an input device such as a graphics tablet, an alphanumeric terminal, and some disk storage. It would often be desirable also to add some form of hardcopy unit, and a local area network interface. Such workstations currently cost between 25% and 100% of the annual salary of a design engineer, and are likely to become cheaper with time.

A local area network interface can provide a high speed link (of the order of 1-10 megabits/second) to other processors. Thus, tasks can be dispatched to the processor most suited to execute them. For example, the solution of a large equation system may be best done on a large processor (e.g., a supercomputer) attached to the network whilst the initial data preparation, and subsequent result interpretation, may be more appropriately handled by a smaller local processor equipped with good display hardware.

GRAPHIC INPUT AND OUTPUT

A wide variety of graphic devices has been used for modelling, analysis, and design, differing in their display abilities. Similarly, various input devices now exist, with different characteristics.

Graphic Displays for CAD

Early graphics terminals were either of the pen and ink variety, or the storage tube kind. Both are capable of producing monochrome pictures of very high quality. Both are line graphics devices, unable to create halftone pictures.

The storage tube writes displays on a phosphorescent screen with an electron beam, exactly as a television picture tube does. But the phosphor is cleverly arranged so that the image remains even after the electron beam has been removed; it is "stored" by the screen material itself. Storage-tube displays are static, exactly like pen-and-ink plotters: once a line has been drawn, it remains on the screen until the entire screen is erased.

Raster displays are based on television technology. The image only remains on the screen for a some milliseconds, and has to be refreshed continuously, usually either 30 or 60 times per second. The picture is not stored by the screen itself, but is remembered by a semiconductor memory similar to computer memory. One memory bit per picture dot (pixel) is the minimum required; colour pictures of course need more. The picture definition, which is limited by television technology rather than by the cost of memory, is not quite so good as that of a storage tube, but halftone or colour images are possible. Most importantly, raster displays are dynamic: since they permit any part of the picture to be erased or replaced in milliseconds, a picture can be altered or edited without erasing it.

The importance for CAD of dynamic displays cannot be overemphasized. Successive displays in CAD often do not differ much from each other; with a dynamic device, it suffices to transmit changes in the displayed picture, not to erase and slowly redraw the entire display. Graphic working speeds are much higher than with a static device, for much redundancy in communicaton is eliminated. As a result, the user communicates with the system in a totally different, and much more flexible, way.

Graphic Input Devices

Graphic input in CAD systems takes two basic forms: commands and data. Commands are often passed from user to system through the graphic display -- the user identifies elements of the display, or selects choices from menus, by pointing at them. This type of input requires sufficiently rapid response to allow repositioning a cross-hair cursor or other marker on the screen every 15-20 milliseconds, so as to follow hand movement. Extreme precision is not required, since the device is generally positioned freehand. Graphic data input occurs in applications such as the digitizing of curves or graphs, where essentially numeric information must be transmitted. Precision is essential in this case, but speed is less critical.

The graphic input device of choice today, incorporated into the

194

majority of new CAD systems, is the digitizer tablet. The tablet is used with a stylus resembling an ordinary pencil, though some workers prefer a "puck" which lies flat on the tablet. Either device contains a small sensing coil which determines its position on the tablet, and communicates it to the outside world through a cable. A position signal is sent out every few milliseconds, so that the stylus can be easily used to follow ordinary manual sketching. The most common varieties of tablet have an active surface about 30 cm square, a little larger than an ordinary writing pad, with a spatial resolution better than 0.2 mm. Larger sizes exist, up to about two meters square. They are sometimes used for digitizing existing drawings or maps. The tablet has won wide acceptance because its precision is sufficient for digitizing graphs while its speed is quite high enough for cursor steering.

Other graphic input devices encountered occasionally include thumb-wheels, trackballs, mice and joysticks. None of these is usable for numerical digitizing. They all work by means of two potentiometers which detect x and y positions independently. Thumbwheels have been used for years in storage-tube terminals. They have the advantage, much liked by some, of permitting cursor movement in the x and y directions independently. In fact, they are the only device with which it is possible to move the cursor along a true vertical or true horizontal. Trackballs are tennis-ball sized shperes placed in a suitable box, riding on two potentiometers. Mice are objects which can be moved about on a flat surface, they are essentially trackballs turned upside down. Joysticks come in two varieties, position controlling and velocity controlling. Position control joysticks are quite agreeable, but cost at least as much as a graphics tablet, since analogue to digital conversion with at least nine or ten bit precision is required. Velocity control joysticks are cheap to make, but inconvenient to use. They have been so roundly condemned by the CAD user community that very few systems even offer them as options.

Early refresh graphics devices used the lightpen as an input device. In principle, the lightpen seems attractive; in practice, it is not. Because it is used to point directly at the display screen, it requires working on a vertical surface, which is tiring. Alternatively, the display tube must be mounted horizontally, so the operator has to work at a table with no kneeroom at all. Sufficient positional accuracy can be achieved with lightpens only at the blue end of the spectrum, a matter of no great consequence in monochrome but rather disturbing in colour work where every lightpen hit is accompanied by a blue flash of the screen. Lightpens cannot be used to digitise existing drawings, cost as much as tablets, and seem to offer no compensating advantages.

Speeding the Graphics Terminal
A picture is proverbially said to be worth a thousand words, an estimate is likely to err on the low side. Only a drawing of rather

modest complexity can be encoded and transmitted to a graphics terminal as a string of a few thousand characters. Consequently, graphics work is limited by data transmission speed. Alternatively, high working speeds can be achieved by one of two techniques: use of distributed intelligence, and use of dedicated computers.

In systems said to have distributed intelligence, part of the processing power of the computer is moved to the far end of the communications line. Drawings can then be encoded very compactly, in relatively complicated forms, and reprocessed at the terminal end into simpler commands comprehensible to the graphics hardware. The effective communication speed is increased, by making fewer transmitted characters carry more picture information.

When dedicated computers are used, the graphics device is moved into the computer proper, and made to reside on the main computer bus. The bus speed is generally about 1000 times the best obtainable on a dedicated serial communications line, and up to 100000 times that obtainable on a telephone line. Speed is increased in this case by simply eliminating the slowest part of the communication link. This approach has traditionally been part of the minicomputer world, but under the pressure of CAD requirements, even large mainframe machines are now being equipped with graphics devices directly attached to one of the internal buses. In the IBM dialect of English, these devices are referred to as channel-connected graphics units.

Graphics and Distributed Computing
For the reasons already adduced, distributed computing, in which two or machines of differing characteristics are interconnected, is likely to become conventional for CAD. Fig. 1 shows the general layout of a distributed computing system, consisting of a host computer and an intelligent terminal. The host central processor and memory, at the left side of the sketch, are tied together by the host computer interconnection bus. On this bus, information typically moves at a rate of 5 million bytes per second or faster. The display processor (terminal or local processor) and its local memory are similarly connected, though the processing speeds and data transfer rates there may be slower by perhaps one order of magnitude. The two communications interfaces are connected by a serial line capable of moving 30 - 960 bytes per second. This serial connection, which moves data at rates some four magnitudes slower than the host and terminal interconnection buses, forms the true bottleneck.

The classical centralised computer facility of the nineteen-seventies is identical to the system of Fig. 1, but with very little intelligence at the terminal end. Typically, terminals employed in the seventies contained internal memories of a few words, and could understand perhaps half a dozen commands: turn beam on or off, switch from alphabetic to graphic mode, draw a line. The display processor to do so might contain a few chips to implement the

196

necessary rudimentary hard-wired logic. While Fig. 1 still gives a
valid description in principle, the left side of the diagram would
now contain virtually all the processing and nearly all the equipment
used. On the right, only the communication interface and the graphic
display would exist in more than vestigial form.

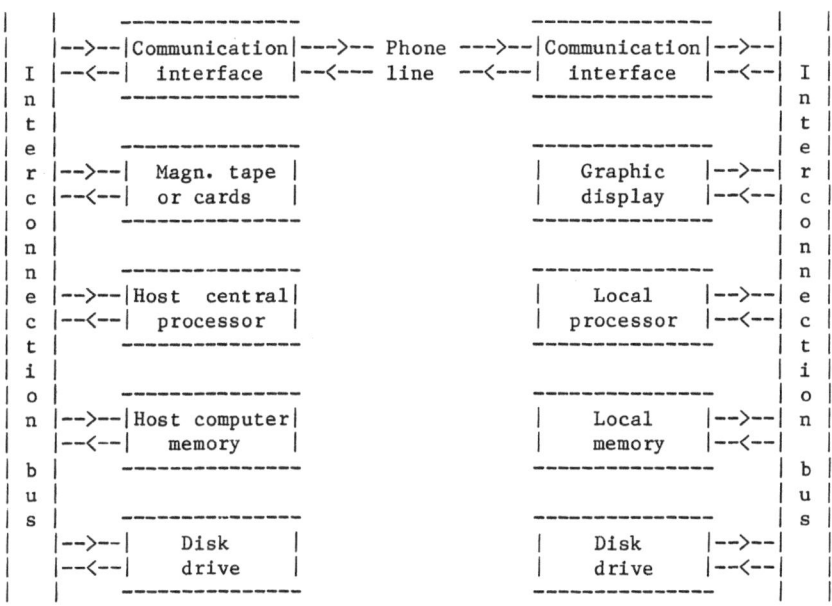

Fig. 1. Hardware structure of a distributed CAD system. Current
systems differ in the relative sizes of host and terminal facilities.

The dedicated minicomputer facility is also in principle similar to
the system of Fig. 1, but tilted to the opposite extreme: the
"central" processor has become vestigial, and the terminal processor
assumes virtually all computing functions. This situation actually
occurs quite frequently; for example, all input/output data
preparation and all postprocessing may be done on the local terminal,
leaving only the equation-solving task for the central processor.
The latter thus handles only one computing process, albeit a very
intensive and demanding one.

ERGONOMICS OF CAD

Ergonomic considerations, including matters so diverse as form of command language and placement of the display screen, are critical in determining user acceptance of entire systems.

Menu Communication

Menus are generally used in conjunction with a graphics screen, on which the items available for selection are presented as outlined screen areas. The user identifies a menu selection by whatever form of graphic pointing his hardware system allows. Within each menu, all those (and only those) selections are presented which the system designer wishes to permit at a given moment. Menu communication does not actually require graphics, it can be just as easily arranged with an alphabetic terminal. For example, the query

Length measured in MILEs, FURLongs, YARDs, FEET, or INCHes? _

initiates a menu selection process just as well as placing the same five choices on the graphics screen would have.

Menu communication is often tree structured, so that choosing a menu item often results in the appearance of another menu on the screen. Hierarchical menus can become a user's nightmare, and are best handled by allowing only the currently queried menu to appear on the screen; otherwise how is the user to know which menu to use? Each menu should therefore be erased from the screen as soon as it a selection has been made. But this criterion makes the menu unsuitable for use with static displays (storage tubes), where selective erasure is impossible.

Screen menus use up precious screen space. The total information content of a graphics screen is fixed by its refresh memory, and any screen space reserved for menus is effectively subtracted from space available for graphics. One solution to this problem is to reserve no menu strip or menu area on the screen, but to use any arbitrary part of the display screen for the menu and to restore its initial state subsequently. Again, static display hardware is unsuitable.

Two-Screen Working

Many users of CAD systems prefer separate graphics alphabetic display screens, so that the graphic display is not cluttered with commands and system messages. Working with two screens requires appropriate work station layout, as well as software systems which respect the needs of two-screen working. Both requirements are primarily ergonomic, and both are tricky to meet satisfactorily.

Most people seem to like an eye-to-screen distance such that the screen subtends a lateral angle of vision around 20-25 degrees. Two screens side by side, each at the appropriate viewing distance for a

25-degree angle of view, require turning the direction of vision (combined neck and eye movement) through about 35-40 degrees, an inconveniently large lateral movement. The problem is at least partially curable by placing the graphics screen above the alphabetic screen. An upward eye movement of some 20 or 25 degrees seems to be less bothersome than the corresponding lateral movement. Movement is also reduced, since display screens are wider than they are high.

Cursor steering with a graphics tablet, using a stylus or puck, is a very natural movement. Indeed when using a stylus it is easy to forget that one is not actually writing on the screen -- provided the tablet and screen are well aligned. The tablet top and bottom must be nearly parallel to the horizontal edges of the display screen, misalignment makes hand-eye coordination very difficult. It is helpful to have the tablet directly in front of the display screen, and to have the screen and tablet subtend the same angular width.

In general, there are two unalterable leading dimensions to a work station. The width of the keyboard which is dictated by the size of the human hand, is always about 35 - 40 cm (though wider ones exist). The size of the graphics tablet is usually about 35 cm square, to accommodate conventional paper sizes, which are also related to the size of the human hand. The work station thus has to have a relatively inflexible width of about 75 cm or wider, with the keyboard and tablet side by side, and next to their corresponding screens.

The width of a normal graphics work station is large enough to make continual switching from keyboard to tablet and back inconvenient. Good software design therefore requires the user to move his hands from keyboard to graphics device, and his eyes from one screen to the other, as rarely as possible. For example, where the system designer has a choice between a keyboard query and a menu, he should choose to employ the device likely to have been used most recently. Thus questions interspersed in strings of graphic input should usually expect replies via menu selections on the graphics screen, while questions embedded in numeric input should require keyboard replies. Which device is active, should be evident at all times by glancing at the screens. The graphics cursor, and menus on the graphic screen, should appear only if graphic input is actually expected, and conversely the alphabetic cursor should appear on the screen only when input is awaited. Above all, the situation should never arise where the user and the system both wait for the other to speak first!

Design considerations of importance include such apparently mundane matters as the height of the screen and keyboard. It is surprising how many CAD system users develop physical tiredness and "computer fatigue", simply because their terminal keyboards are installed on tables designed for writing. Computing may not always cause a pain in the lower back -- but having the keyboard three inches too high above chair level is almost guaranteed to do so!

Colour or Monochrome?

Colour displays are indeed very attractive, and they really do convey more information than monochrome, provided the underlying software makes good use of colour. In overall system cost, colour comes perhaps 10% higher than black and white, thus meriting serious consideration. The main unsolved problem with colour is paper copy.

Paper copies of monochrome displays are readily available from electrophotographic hard-copy units. For colour displays, three technologies are available: photographic reproduction, multicolour ribbon printing, or ink-jet printing. Colour slides are cheap, and a superb medium for presentations, meetings, and conferences, but not convenient inclusion in a report. Colour prints come at too high a price; so do ink-jet printers. There exist cheap dot-matrix printers with three-colour ribbons, capable of printing out a 512 x 512 screen image in about a minute or two. Thus copies are still slower to obtain than in black and white, but the gap is acceptable.

Reproduction of colour hard-copy is a difficult problem. Many engineering reports are circulated to five, ten, or fifty people, numbers too small to justify colour separation printing processes yet too large to print duplicate originals. Colour xerography, like the light-pen, has long been heralded as the technology of tomorrow; but in most offices tomorrow has yet to come.

CONCLUSIONS

The complexity of electromagnetics software, and the rapidity with which computer hardware is evolving, combine to make software packages have much longer prospective use lifetimes than their underlying hardware. Software designers therefore must take steps to avoid too early obsolescence. Such steps could include at least

(1) Technological forecasting, to assess what new types of peripheral devices may be available in five years or so.

(2) Reliance on portable operating systems to minimise dependence on variations in essentially standard (processor) hardware.

(3) Adoption of standard software protocols to the largest extent possible, to simplify software tranport in the near future.

Efforts to improve user communication and user training are needed to broaden the user community, and thereby to make shorter software life-cycles economically viable. However, it is not clear how such efforts are best to be undertaken, at least in the short run.

TECHNIQUES OF POST-PROCESSING
FOR ELECTROMAGNETIC FIELD SOLUTIONS

M. L. Barton, I. A. Ince and J. J. Oravec

Westinghouse Electric Corporation, R&D Center
Pittsburgh, PA 15235 USA

1. ABSTRACT

The key to the value of any electromagnetic field solution is the ease
with which the user can extract from that solution the desired engineering quantities. Described here are some of the techniques used in
the post-processor WEPOSTS for the efficient calculation, manipulation
and display of the results of finite element solutions to electromagnetic field problems. WEPOSTS is an interactive, graphics based post-
processing system designed to give the user maximum control over the
extraction of useful information from a calculated numerical field
solution. Techniques are explained for plotting flux lines, loss and
force density levels, saturation levels and for the user-directed
calculation of derivative-based quantities (e.g., flux density) at
points, over elements, along lines and arcs, and in regions. Several
examples are provided to illustrate the power of these techniques.

2. WEPOSTS OVERVIEW

The authors' philosophy of post-processing is that the user should
have the greatest possible control over the information generated
about the numerical field solution. To achieve this flexibility
without making the system so complicated that it is for "experts
only", a natural command language with sensible defaults and readily
available help and teach facilities is required.

WEPOSTS post-processor is the design tool interface between the de-
signer and design related electromagnetic entities. The graphics of
WEPOSTS is based on the Graphics Compatibility System (GCS). Hence,
WEPOSTS runs on terminals supported by GCS.

The structure of WEPOSTS can be seen in Figure 1. It has four
modules: the graphic interaction and analysis, electromagnetic
analysis, plot control and miscellaneous command modules. A brief
description of the graphics interaction module will be given in the
next few paragraphs. This will be followed by a brief description of
the plot control module. However, the predominant discussion will be
in the area of electromagnetic analysis.

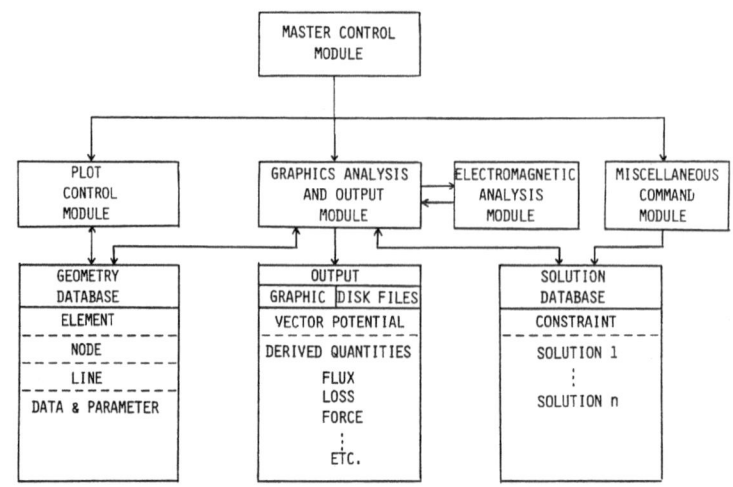

Figure 1.　Post-Processor Structure

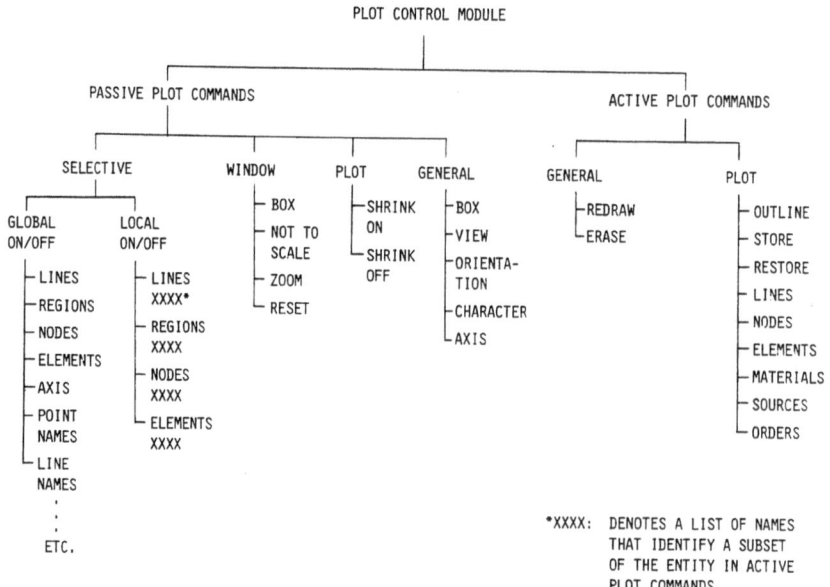

Figure 2.　WEPOSTS Plot Commands

WEPOSTS utilizes two databases: the geometry database and the solution database. The geometry database is generated by the geometric modules (pre-processors). The solution database is generated by one of various electromagnetic analysis programs. Both databases are binary disk files transparent to the user. To enhance the speed of interaction, databases are used only for data retrieval and for user inquiries.

For maximum user friendliness and acceptance, the user interface conforms to a designer's natural path of thinking. All communications are done in terms of engineering design quantities. The user has the option of communicating with WEPOSTS either by the use of menus or through keyboard command strings.

As part of its user interface, WEPOSTS has extensive on-line documentation. At present three levels of on-line documentation are implemented: 1) echo, 2) help, and 3) teach. Entry of HELP followed by an empty carriage return echoes back all the valid command available to the user. The echo feature echoes the arguments of a valid command upon its entry. TEACH is the most extensive on-line documentation and takes the novice user through the commands and functions of WEPOSTS.

WEPOSTS command structure is designed to conform to a tree structure. At every level there are a fixed, pre-programmed set of commands to be allowed. Each command node may end up in another branch or in a leaf (path termination). At every level there are a set of pre-defined defaults which the user has the option of accepting or changing.

Upon user request, WEPOSTS generates output in the form of graphical displays and/or screen tabulations. Formatted disk files for archiving are also provided at the user's requests.

Plot Control Module
WEPOSTS provides the user with extensive control and flexibility in model plotting. A high degree of selectivity has been deliberately implemented. This results in a large set of plot control commands. A subset of the available commands is given in Figure 2. As can be seen, plot control commands are divided into two groups: the "active" and the "passive" plot commands.

Active plot commands are executed immediately after the command is given. These commands may be used to plot a single entity or a group of entities. They may be used to add selective additional information after a global model plot has been completed.

Passive plot commands control what is to be plotted upon entry of the next REDRAW. As implied by their name, the selective plot commands control plotting of model components as well as item names.

3. PLOTTING TECHNIQUES

Although few users would identify the isoline plot as the primary goal of their analysis, such a plot showing flux lines (for magnetic field problems) or equipotential lines (for electric field problems) gives the user an immediate and satisfying qualitative feedback on the success or failure of his model and its calculated solution. In the authors' experience the isoline plot also serves the important function of helping the user communicate the nature and success of his project to others involved in the design process, and gives added weight to the numerical results (e.g., losses) that he may also be presenting.

To achieve these goals a substantial effort has been directed toward the generation of fast, accurate and visually pleasing isoline plots. WEPOSTS determines and displays problem and interface outlines, plots isolines for flux, equipotentials, loss and force densities, and saturation levels. It also displays these various quantities in colo when color graphics devices are available. Because prompt, inter-active response is essential, WEPOSTS maintains a list in active com-puter memory of those nodes and elements currently on screen. Only these entities need be considered during execution of subsequent commands. Through the use of ZOOM and WINDOW commands the user can keep only the area of interest on the screen, thus improving both the resolution of model definition and the efficiency of post-processor execution.

Outline Generation and Plotting
When the finite element system user operates the post-processor to examine his solution, his approach is usually that of an engineer who thinks about his problem in terms of its component parts rather than in terms of finite elements or model building components such as construction lines. WEPOSTS cooperates with the user who takes this approach by displaying the problem region in terms of these component parts as determined by the model's source and material interfaces.

It is important to display these component parts, these material and source interfaces, as determined by the solution database rather than by the user's region definitions as made in the pre-processor. The user need not build his model up from regions which constitute the distinct component interfaces of the problem, this being merely one way to construct models. And even if it could be assumed that the user intended to identify component parts as distinct regions, such a display would provide no additional feedback concerning what problem the system actually solved.

To give the user a complete picture of where the source and material interfaces lie the post-processor must examine the database to determine across which elements the material or source labels differ. All such element edges can then be plotted, preferably in some way distinct from the flux lines (e.g., dashed lines or different colors) and the result will be a plot of all the subregions which differ in material or source label.

Some systems perform this determination by comparing each side of each element with each side of every other element. If N is the number of elements and M the number of sides per element, the total number of comparisons which must be made is the number of permutations of MN things taken two at a time. This number is

$$C = \frac{MN \ (MN - 1)}{2} \tag{1}$$

For a problem with 4000 triangular elements, the number of comparisons, C, is nearly 72 million. Clearly this is not satisfactory.

The outline generation algorithm used in WEPOSTS examines each element only once so that the work required grows linearly with the grid size rather than as the square as in Equation (1). The method starts by taking a fixed amount of memory (the penalty we pay for working in FORTRAN) and apportioning it into

$$(S - 1) + (M - 1) + 1 = S + M - 1 \tag{2}$$

equal parts where S is the number of distinct source labels and M the number of distinct material labels. This number of parts is arrived at by observing that the problem boundary must be drawn, then (S - 1) source and (M - 1) material subregions must be drawn with the last source and material subregion each coinciding with either the problem boundary or the boundary of some adjacent source or material region.

The algorithm now proceeds by examining each element to determine its source and material labels. The node pairs constituting the sides of the element are added to the list of node pairs for that source and material (unless they belong to the single excluded source or material) and to the list of node pairs for the problem boundary according to the following rule: If the node pair is not in the list, add it; if it is already in the list, delete it. At the completion of this process, only those element edges which have been found only once in conjunction with a particular label will remain in the list, and these are exactly the edges whose union constitutes the outline for that label.

205

Contour Plotting

Contour plotting is available for flux lines, loss and force density distributions and saturation levels. The plotting algorithm captures the accuracy of the calculated solution in both high-order triangular (Konrad and Silvester, 1973) and dual order parallelogram (Barton, 1982) elements. Plotting options include VALU-NOVALU for an optional list of plotted contour values and FAST-SLOW for a choice between a FAST (and good) plot and a SLOW (and very good) plot. Figure 3 shows the generated outline and flux contour plot for an axisymmetric silicon crystal growth facility. In this device the coil and yoke are used to produce a uniform magnetic field through the silicon melt.

4. CALCULATION TECHNIQUES

Two-dimensional finite element analysis programs for electromagnetics almost invariably calculate a potential (scalar for electric or vector for magnetic problems) rather than a direct field quantity. Rarely is this the quantity that the system user wants. More commonly he wants the flux density (in a magnetic field problem) or some quantity to be derived from flux density (e.g., force). Since most quantities of engineering interest are based on derivatives of the calculated potentials (e.g., flux densities), the post-processor needs an efficient real-time evaluator of these derivatives. It would not be unfair to say that the user's acceptance of the calculation module in the post-processor depends primarily on how fast derivatives of finite element solutions can be calculated, manipulated and displayed. WEPOSTS employs differentiation matrices as in Csendes (1975) in two dimensions and performs all flux density related calculations on an element by element basis as required for a particular user request.

Derivatives in Triangular Elements

An n'th order polynomial function, u, in a triangular finite element can be written as

$$u\ (x,y) = \overline{\alpha}^n\ \underline{u} \tag{3}$$

where $\overline{\alpha}$ is a row vector of interpolation functions and $\underline{u}$ is a column vector of nodal values. The derivative of the n'th order polynomial function, u, can be exactly represented by a complete polynomial function of order (n - 1). The vector of coefficients of the (n - 1) order polynomial functions, v, describing a derivative component is

$$\underline{v} = D^{n-1,n}\underline{u} \tag{4}$$

where $D^{n-1,n}$ is a differentiation matrix of dimension k x l and

$$k = \frac{n\ *\ (n + 1)}{2} \tag{5}$$

206

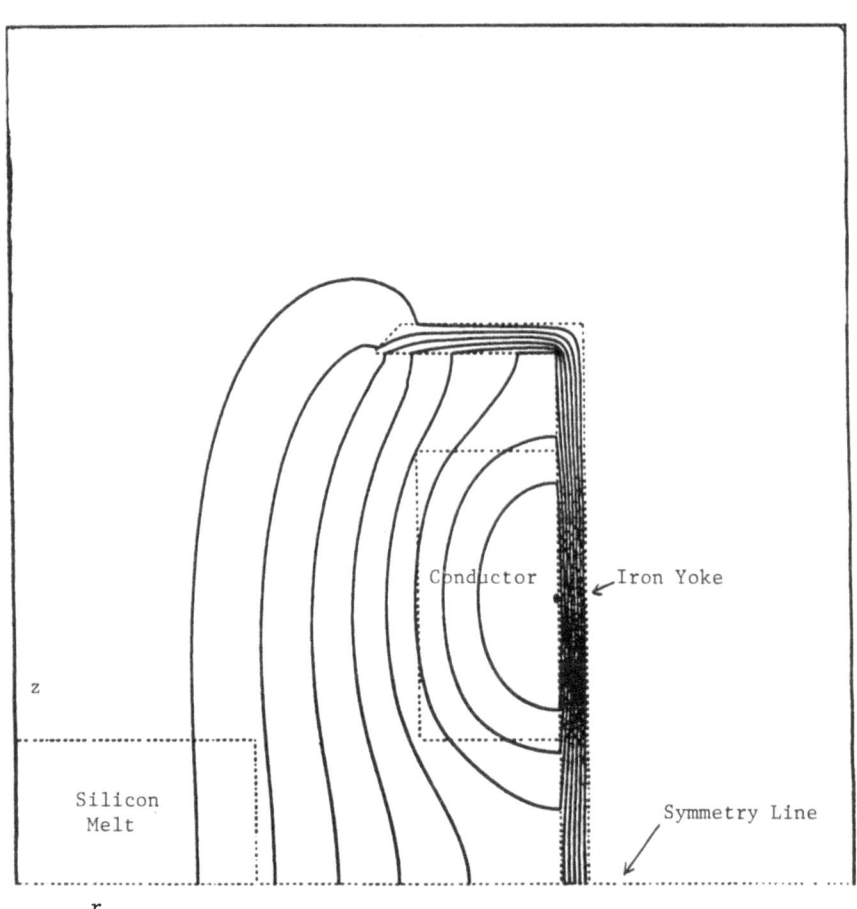

Figure 3.   Outline and Flux Plot for Axisymmetric Silicon
Crystal Growth Facility

and

$$1 = \frac{(n + 1) * (n + 2)}{2} \tag{6}$$

are the number of nodes in an (n - 1) order and n'th order triangle respectively.

The entries in the matrix D are given by

$$D_{ij} = \sum_{K=1}^{3} C_K G_{ij}^K \tag{7}$$

where the quantity C_K is a geometrical constant and the matrices G^K may be calculated once and for all. Since the above expressions are assembled from the product of universal matrices and geometric constants determined from the individual triangles, generation of differentiation matrices is quick and inexpensive at run time.

The above definition of a rectangular differentiation matrix is very useful in applications where the user wants to see the results of some calculation involving derivatives, or where he wants to see derivatives at arbitrary points on the mesh. If, on the other hand, he wants to see derivatives at the nodes of the elements, as he often will, a more efficient implementation is available. Since the derivative function on an element is of order (n - 1) it can be exactly modelled by any complete polynomial of order (n - 1) or larger. Thus the derivative function can be automatically interpolated onto the original n'th order element nodes by evaluating the square l x l matrix

$$H_{ij} = \left. \frac{d \alpha_j}{dx} \right|_{(x_{ij} \ y_{ij})} \tag{8}$$

where both indices i and j extend over the n'th order nodes. Maximum flexibility is afforded if both the rectangular and the square differentiation matrices are available to be employed according to the instructions of the user. Y-directed derivatives are handled in an analagous way.

Derivatives in Parallelogram Elements

For dual-order parallelogram elements (Barton, 1982) the interpolation functions from Equation (3) can be written

$$\alpha_i = \beta_\ell(u) \ \beta_m(v) \tag{9}$$

where u and v are normalized coordinates related to cartesian coordinates by

208

$$
\begin{bmatrix} u \\ v \end{bmatrix} = \begin{bmatrix} a_1 & b_1 & c_1 \\ a_2 & b_2 & c_2 \end{bmatrix} \begin{bmatrix} x \\ y \\ 1 \end{bmatrix} \tag{10}
$$

and the coefficients a, b and c are determined from the vertex locations of a particular element. From Equations (9) and (10) the
derivative of an interpolation function may be expressed as

$$
\frac{\partial \alpha_i}{\partial x} = a_1 \frac{d\beta_\ell}{du} \beta_m + a_2 \beta_\ell \frac{d\beta_m}{dv} \tag{11}
$$

The full derivative quantities in Equation (11) are universal expressions unrelated to individual elements and can thus be calculated
once and for all and stored. This is particularly simple for paral-
lelograms since the quantities to be calculated are strictly one-
dimensional.

Displaying Derivative Quantities

The post-processor provides a variety of mechanisms by which information about the derivatives may be obtained. In addition to the plots
of equal levels of saturation mentioned above, the user may request
flux density components at nodes or over elements. These requests
are made simply by imputting the command FLUX NODE or FLUX ELEMENT
and pointing to the appropriate screen location with the cursor.
FLUX NODE returns the weighted average value of flux density at the
finite element node nearest to the cursor. FLUX ELEMENT returns the
flux density components at the centroid and the average value (when
they differ).

The commands FLUX LINE and FLUX ARC allow the user to see the variation in flux density along a specified line or arc. Again the cursor
is used to specify line endpoints or arc centers and angle spans.
The result of either of these commands is a pair of graphs giving
flux density components as functions of normalized distance along the
line or arc. The user controls whether the output is in x-y components (or r-z or r-theta as appropriate) or normal-tangential components or magnitude-phase. Figure 4 shows a flux plot for the upper
half of a "C"-shaped core section driven by a rectangular conductor.
The solution is symmetric about the bottom boundary. The line drawn
from the "+" sign is the requested line for flux density calculations.
Figure 5 shows the resulting graphs.

In Figure 6, the flux plot for one quarter of a motor cross-section
is shown. The arc drawn from the "+" sign is the requested arc for
flux density evaluation and, since this problem is defined in r-θ

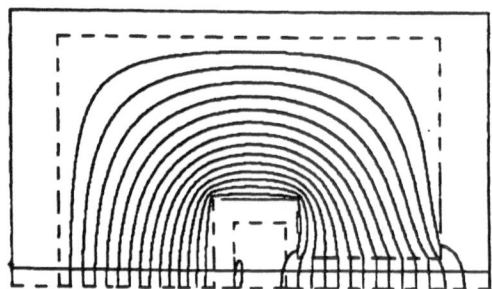

Figure 4. Flux Plot for 'C'-Core

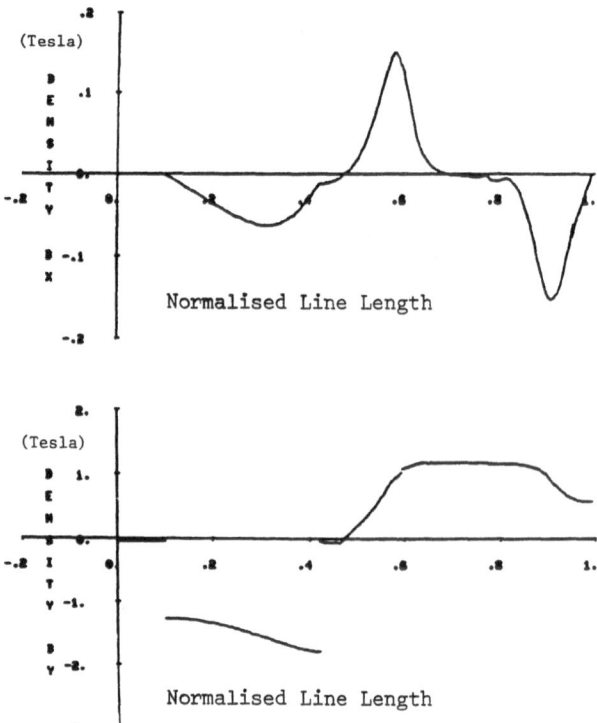

Figure 5. Flux Density Profiles for 'C'-Core

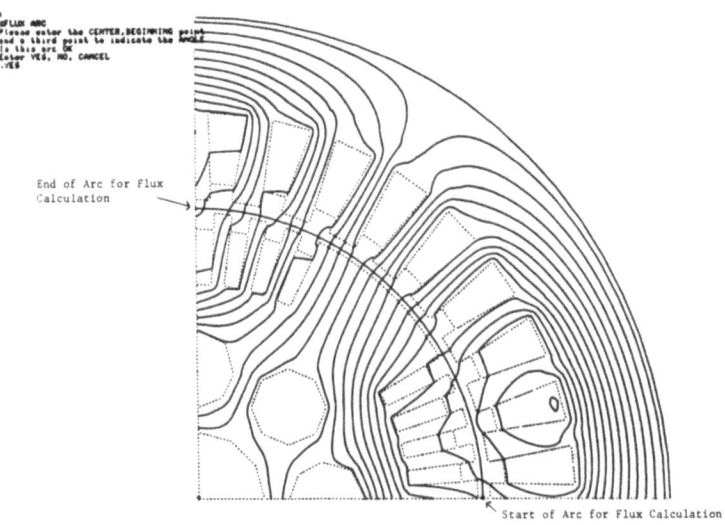

Figure 6.   Flux Plot for Motor Cross Section

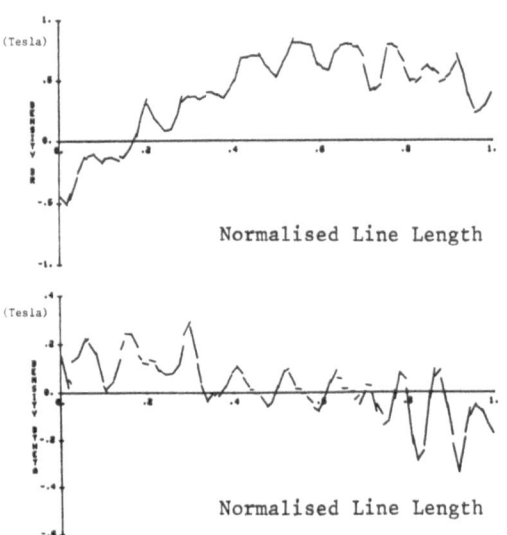

Figure 7.   Flux Density Profiles for Motor

211

coordinates, the r-θ flux density components will be evaluated. The graphs of these airgap flux density profiles are shown in Figure 7.

In Figures 5 and 7 the flux density values have not been smoothed as they cross element boundaries. A procedure to average the discontinuous values of flux density across these boundaries is available, but its use would obscure what the authors find to be an excellent indication of the accuracy of the solution -- when the derivatives of the potential become nearly continuous, the solution is a very good one.

Other functions of the post-processor include determination of maximum flux density over the problem region, material region or user-defined windows; calculation of energy related quantities such as inductance; and determination of total force or torque on bodies.

5. SUMMARY

WEPOSTS described above is an interactive graphics post-processor for finite element analysis programs. It is tailored to the needs of the electrical designer. By doing all its interaction in terms of standard design terminology, it enables the designers to use the computer as a friendly design tool without forcing them to learn computer terminology. The wide range of electromagnetic entities derived by WEPOSTS and its selectivity enables the user to concentrate on the problem at hand without having to go through reams of computer output. The interactivity and the near real time speed of response enables the design of electrical machines interactively. WEPOSTS is a model example of a profitable partnership between the designer and the computer in effectively dealing with complex design problems on the road to better performance and efficiency.

References

Barton, M., "Dual-Order Parallelogram Finite Elements for the Axisymmetric Vector Poisson Equation", IEEE Trans. on Mag., Vol. MAG-18, No. 2., March 1982, pp. 599-604.

Csendes, Z., "A Fortran Program to Generate Finite Difference Formulas", Int. J. Num. Meth. In Engng., Vol. 9, pp. 581-599, 1975.

Konrad A. and Silvester, P., "Triangular Finite Elements for the Generalized Bessel Equation of Order m", Int. J. Num. Meth. in Engng., Vol. 7, pp. 43-55, 1973.

COMPUTER DEVELOPMENT AT SUNDERLAND POLYTECHNIC

- HAS IT PROVED TO BE A PERIODIC FUNCTION?

E Crompton and J Tindle

Department of Electrical, Electronic and
Control Engineering, Sunderland Polytechnic

1.1 INTRODUCTION

When the first generally available Digital Machines were introduced
into the design process by non-specialist manufacturers in the
1960's electromagnetics were used on these machines in three different ways; these were:

a) The computation of machine transients.
b) The calculation of machine characteristics as an aid to design.
c) The empirical cost estimation of machines for sales tendering.

At the time it seemed that the digital machine would replace most
analogue machines within the next two years and the drawing office in
a decade. Any type of calculation was attempted using the first high
level languages and when space problems arose on the first mainframe
machines machine code inserts were written to overcome this problem.
A typical computer at this time had a very limited direct access
magnetic core store, a magnetic drum backing store, a single paper
tape input channel, and a line printer output. Programs had to be
written which involved frequent disc transfers of instructions and
data and care had to be taken not to exceed the expected operational
life of the machine; very long processes may never be completed
because of a system breakdown.

Since that time even Polytechnics have had as many as ten main frame
machine replacements and at least six new high level languages each
offering a partial solution to computational problems with a substantial overhead in recoding at almost every stage.

The languages were: Autocode, Fortran, Algol, Focal, Apl and Basic.
In 1981/82 Sunderland Polytechnic purchased two multiple access
super-mini computer systems to:

1) Provide a CAD/CAM facility for the Polytechnic.
2) Update an existing service used for undergraduate teaching.

After all the feasibility studies and benchmarking processes had
been completed the personnel involved wondered if this was the last
time it would be necessary to purchase this type of multi-access

machine. One of these machines has now been in operation for 6
months and the other for about 1 year and the Polytechnic staff have
experienced a software change-over without an excessive reprogramming
overhead for the first time.

2.1 PROGRAMMING LANGUAGES

On any computer the combination of programmer and external designer
using the equipment will naturally develop an interactive analysis
system to meet their own needs whatever type of high level language
is used. In any case only a minimum sub-set of the instructions
within the language specification will be used at any one time.
This working vocabulary evolves as the personnel gain experience and
will be a function of the current "fashion" of the people concerned.
Departments or individuals with a large software library will be
very resistant to the introduction of a new programming language
unless the system is specifically tailored for reliable mixed language coding.

It may be that the most receptive personnel to language changes are
those who have not progressed very far in the development of complex
packages, or groups which have an obvious interest in this type of
work. For example, at this time, it can be said by a number of
researchers that PASCAL is not yet sufficiently developed to warrant
recommending its use for new software in magnetics at Sunderland
Polytechnic, while others prefer to use the language for all their
work. The operating systems available at Sunderland do not have a
robust FORTRAN to PASCAL interface and at the moment the increased
power of PASCAL in some data handling modes cannot be used.

It will be interesting to see if FORTRAN is modified by the influence
of PASCAL to kill its development as happened to ALGOL in the past.
Even with an internationally accepted standard for a language the
tendency is for the manufacturer to extend the code to maintain his
competitiveness. This characteristic results in many traps for the
unwary because these extensions often provide very attractive
options for the programmer to use. The result is non-transparent
software, e.g:

In FORTRAN-77, character data always includes inserted blanks, these
are used to maintain the specified string length in a Character
Variable. This can cause problems when string characters are being
compared or concatenated. On Harris compilers a *(*) character
function can be written to remove the blanks in a general manner;
this function cannot be used on the VAX series of computers.

Other HARRIS/VAX discrepancies occur in the looping instructions
available on one machine which differ from those on the other -

One compiler allows BASIC like loops of the type

```
FOR.........(CR)          DO
*                         *
END FOR                   UNTIL (cond. clause.)
```

While the other uses

```
DO..........(CR)          DO WHILE (cond. clause.)
*                         *
END DO                    END DO
```

Our departmental experience with the various languages available to
users on the Polytechnic computers is summarised in Table 1 shown
below.

Computer	Language	Available	Comments
HARRIS	FORTRAN-77	yes	Good compiler with many non-standard Features.
HARRIS	PASCAL	yes	Rather fragile compiler not recommended for general use.
VAX 750	FORTRAN-77	yes	Compiler has less non-standard Features than the Harris. There are problems with dummy variables in Subroutines.
VAX 750	PASCAL	yes	Compiler used extensively. Mixed coding from PASCAL to FORTRAN has been used.

Table 1 Comments on High Level Languages

Basic has not been included in Table 1 because it is not a suitable
language for the type of computation considered. Basic is usually
available on all computers ranging from micro's to mainline machines.
In some ways the character handling functions in this language are
superior to those of FORTRAN-77.

At the moment there are no universally accepted standard graphics
languages implemented for multi-access machines although licences
are easily purchased, at a price, for a number of commercial
packages. In addition industry seems to favour turnkey graphics
systems which are out of the price range of an Educational establishment.

215

Low cost Microcomputers are being marketed which have the apparent capacity of a shared machine to a single user with the advantage of portability. The size, cost and specification of these units will be considered in another section of this paper but it is obvious that the unit cost is much less than that for a multi-access machine. The impact of these machines on the user (probably the inexperienced user) and the problems which have arisen in the past which should be avoided in the future will now be discussed.

When considering these micro-systems, which are being developed in many centres simultaneously, the situation seems to be almost parallel to that occuring in the 1960's. The machines now available are vastly more powerful than those initally used, the cost per unit has been divided by a factor of 10 to 100 in this time, but some of the lessons from the past have not yet been taken to heart.

With each new family of machines new command instructions and operating systems are developed and a number of alternative high level languages are offered. These changes could be said to be partially due to scientific improvements but the effect of commercial competitiveness must not be overlooked. From the point of view of the Scientific user not all of the alleged improvements in performance are necessarily productive and some features can limit your hardware choice in the future.

To minimise the cost it is essential that Software must be written to outlast the hardware processing the code. If a program is transparent to a number of computers the operating flexibility is increased and the downward compatibility of existing software is guaranteed. Coding optimisation, essential for the very large complex packages in magnetics, should be carried out by a system option in the machine being used and not by the use of self coded inserts which are very attractive in individual machines. When the amount of time being spent by computer staff simply maintaining the operating system of a multi-access machine is considered the uncontrolled combination of a number of powerful micro's in any establishment is a frightening prospect.

When moves are made from package based data structures to company wide data bases the requirements for downward compatibility become even more stringent. In this case the data may outlive both the software and the hardware used to process it. Information must be designed to last the life of the equipment it describes and the base may have to be referenced urgently at any time in this life. Therefore data which is structured in a similar manner to a program will give a significant saving in processing overheads. This is because all operating systems on a computer are able to access data of this type via the "in house" editor.

On the multi-access machines at Sunderland Polytechnic we have solved the problem of simple standard graphics for undergraduates by writing our own compiler. The Device Independant Graphics system (DIG) uses simple alpha-numeric code to describe standard graphic processes. Handlers are provided on all our machines to convert this code to graphs, either on terminals or hard copy devices. This enables the user to transfer his results to any machine which has the facilities he wants to use.

On the VAX machine this system generates compatible code to the existing GIGI colour graphics editor and produces curves directly from a simple data array. The user does not need to know the type of terminal he is working on, the translation being automatic. Even our teletype terminals change to "line printer" graphics if these sub-routines are linked to any program.

The CAD machine with a greater number of Vector-Graphic terminals has enabled us to develop a number of standard programs to convert output listings of an old CAD package to graphs. The technique of intercept graphics enables the user to link existing batch analysis programs to graphics output and make comparisons between data runs on these packages without changing the existing structure. A series of curves obtained by using this system are shown in Figure 1. The results represent a graphical comparison between one line Voltage in a transient study of a power line surge model as a function of the model complexity. The technique can be used by manufacturers who have a considerable number of batch analysis packages already operating in their production stream without going to the expense of re-programming to convert to graphical output. The only data source required is an old line printer output of the program and some understanding of the information in this listing. This knowledge will be available in the section using the production program. The output of the package must be stored rather than listed and scanned interactively to find the data to be analysed.

3.1 FUTURE DEVELOPMENTS IN ELECTROMAGNETIC COMPUTATIONS

The main problems associated with the changeover to better design facilities using the full potential of interactive graphics are:

Cost
Organisations, which have in house computer service departments, will prefer to modify existing software rather than replace it with a complete system unless the underlying commercial pressures are overwhelming. The Electromagnetic part of design is small when compared with the Mechanical design and the logistic problems involved in manufacture. The introduction of computerised draughting and manufacturing systems will give more obvious commercial rewards

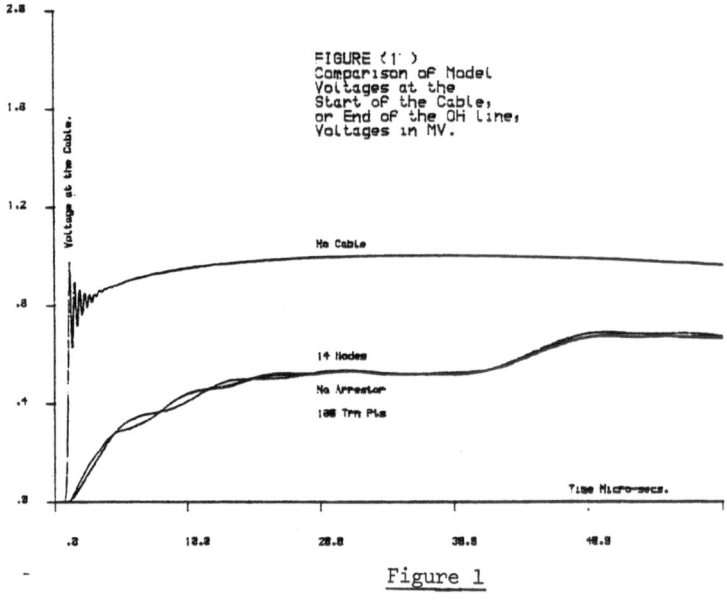

than most electromagnetic analysis packages. The linking of data
bases with these analysis packages may have a low priority particularly if the organisation has other analysis systems already available. A manufacturer cannot make an immediate change over to a new
system in the same way as we can change a car because his production
must be maintained during this period.

Personnel
When computers are introduced as an interactive design tool not all
the company personnel are sympathetic to the unavoidable intrusion of
this device into their lives, either because of the implication that
an increase in efficiency without an increase in orders means a
reduction in staff or because only a limited number of staff can
adapt to the media.

Interpretation
With on-line graphics, in full colour, the options to the designer
are so radically different that the precise alignment of the computational results to the designers' own concept of the machine will
take time. Afterwards the matching of these results to available
test data may be very revealing because of the tendency of the
personnel involved to forget that they are working on a model and not
an actual machine, transformer or coil. The closeness of the
computational fireworks may make the designer forget the theoretical

limitations of the analysis techniques used.

Data Life
A manufacturer is designing equipment to be constructed for an
external customer. This customer may expect the fundamental design
information to be available to his engineers for the life of the
equipment. Therefore most companies are expected to retain parts
listings and drawings of their output for a very long time, particularly in terms of the lifetime of a particular computer. The
digitisation of this information must include in the structure of the
data sufficient text for it to be accessed in the future, even when
the original hardware used to assemble the data file has been
scrapped. In addition more and more customers will expect data
files for their equipment to be made available to them in machine
readable form when it is delivered.

Electromagnetic Finite Element packages in two or three dimensions
seem to be limited to an advanced modelling technique with very
sophisticated pre and post processing programs. The amount of output
involved in true design computations can be illustrated by considering
the computation of the excitation current required for a given a.c.
machine at a given output which is a standard design problem. To
solve this problem a designer using a modelling package will have to
make a number of external "loops" using the program to calculate one
point on the characteristic where the unknowns are the excitation
and load angle for a given fundamental flux density in the air gap
and not the flux density for a given set of currents.

4.1 RECENT DEVELOPMENTS IN MICROCOMPUTER TECHNOLOGY

First generation microcomputers may be considered to be obsolete.
Second generation machines are well established and are now being
superceded by third generation machines. The machines under
consideration have the following features:

Second Generation Microcomputers
Second Generation Microcomputers have the following characteristics;
PROM based interpretive software, limited RAM area and expansion
facilities, offer only a limited pixel graphics capability and have
a limited disc storage facility. Machines of this type are
unsuitable for use in complex engineering CAD/CAM systems. An
example of this type of machine is the Commodore PET.

Third Generation Microcomputers
Third Generation single user microcomputers are compiler oriented,
have multiple high capacity disc drives, significant computing power
and high resolution screens, ref. (1,2,3,14). Machines of this
type available within Sunderland Polytechnic are the Sirius-1,

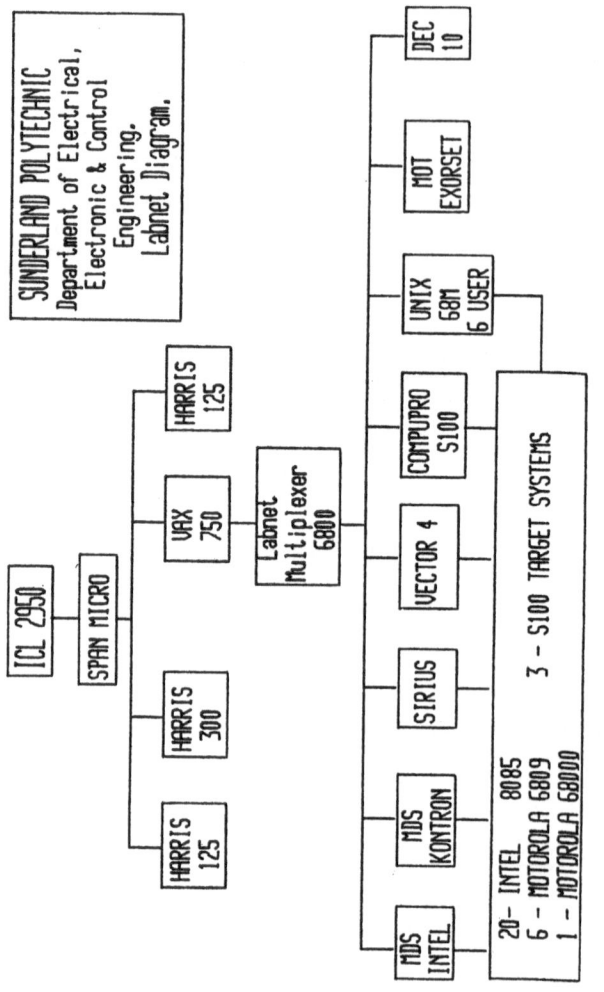

Figure 2

220

Vector-4, and Compupro S100/IEEE 696 based systems. Other machines recently obtained with significant computing capacilities are a PERQ and a multi-user 68000 system running under IDRIS. Figure 2 shows these machines interconnected within the Electrical Engineering Department laboratory network - LABNET. Third generation microcomputers cost between five and twenty per cent of the cost of a typical multi-user mini-computer. A development cycle time for 40K byte of object code is in the order of eight minutes. Compilation times may be reduced by hardware configurations that include Winchester harddiscs, RAM based disc emulators and floppy discs units that use DMA data transfer. Unfortunately, a disadvantage associated with harddiscs is their lack of portability. Program execution time may be reduced by utilising a numerics processor extension, NPX, or a floating point processor unit, FPP.

At Sunderland Polytechnic a twenty fold increase in processing speed has been achieved on the Sirius by linking into Pascal a hardware FPP for matrix manipulation, ref(9). As a further development a parallel port driven graphics tablet has recently been interfaced to the Sirius computer to provide an interactive drawing facility, ref (10). To drive the graphics tablet, machine code software has been developed in-house and linked into Pascal. The principle features of a third generation microcomputer are discussed in the following section.

4.2 THIRD GENERATION MICROCOMPUTER SPECIFICATION

The principle features of a general purpose workstation capable of running MSDOS are:

1) RAM from 128Kb as standard, up to 892Kb maximum, a minimum of 256Kb is required to run Pascal.
2) Reconfigurable keyboard/character set.
3) High resolution graphics screen, typical requirements are: 800 by 400 dots with 8 pages requiring 40Kb per page. A RAM area of 500kbyte should be allowed for graphics related software.
4) Terminal emulation on the computer is essential because it enables the power of minis and mainframes to be accessed by a modern desktop 16 bit microcomputer. File transfer in both directions enables work to be processed locally or on the mini-computer with reports, or updated files, returned to the larger central machine giving full distributed processing capability to the user.
5) The computer must be able to run a high level language such as Pascal. Pascal programs when compiled can be designed to run in 64K code segments, but programs greater than 64K may be created by adjusting segment boundaries. Pascal and Fortran represent real numbers in single precision to the IEEE format, giving a range of $1.10E+38$ to $1.10E-38$, with a precision of seven decimal

221

places. Double precision procedures and functions written in
Basic may be linked into Pascal.

6) A graphics printer to facilitate screen dumps is extremely
   desirable. A proprietary graphics kernel is required which
   occupies typically 20Kbyte.

7) Matrix handling ability; using single precision number representation with 64Kb of memory gives 16K real numbers thereby providing a 90 by 90 matrix.

8) Disc storage capacity:
   2 - 0.6Mb or 1.2Mb floppies or
   1 - 1.2Mb and 1.10Mb hard-disc plus an optional standalone
   20Mb hard-disc.
   Arrays may be placed anywhere in memory by absolute address
   reference pointers.
   An 8 bit 64Kb sub-system may be added giving the facility of
   switching between 8 and 16 bit CPM systems.
   Wordprocessors such as Wordstar are useful when generating HLL
   source programs.

The Sirius, Vector-4 and IBM-PC machines all have similar specific-
ations.

4.3 THE PASCAL PROGRAMMING LANGUAGE

Pascal is now widely used within Sunderland Polytechnic for teaching
purposes and industrial consultancy because it: offers an algorithmic
solution to problems; has expression abstraction, control abstraction
and a certain amount of data abstraction; is available on minicomputers, microcomputers, and on our Kontron MDS; may be used effectively in realtime data gathering and control situations; is structured
and highly readable; has a record mode format that is useful in database applications, ref (11).

The limitations of Pascal are that it does not include as standard
some scientific functions and double precision arithmetic, and also
strong data typing can at times be restrictive.

4.4 A REVIEW OF SINGLE USER MICROCOMPUTER OPERATING SYSTEMS

The Control Program for Microcomputers CPM 80

The most widely used operating system is at present CPM 80 produced
by Digital Research. Great reductions in the cost of standard
software, such as wordprocessors, compilers and business packages,
can be directly attributed to the success of CPM. The CPM 80
operating system runs on 8 bit processors and can be made to run on
a variety of different systems because only the BIOS or basic input
output system is machine dependent, ref. (4,5).

Note. It is unlikely that 8 bit systems will be able to solve all but the simplest finite element magnetics problems because of their limited memory size and numerical precision.

CPM 86

The CPM 86 operating system runs on 16 bit processors and is to some extent compatible with the CPM 80 system as source programs written for 8 bit machines may be translated and made to run in a 16 bit environment. One of the most important features of CPM is that it provides some degree of compatibility between 8 and 16 bit software packages, ref (6,8).

Microsoft Disc Operating Systems MSDOS

Microsoft have produced a 16 bit operating system, called MSDOS which is well established and is now used in most 16 bit applications, ref (7). A number of HLL running under MSDOS have been produced by Microsoft, such as Fortran, Basic and Pascal. The Microsoft MSDOS system is comparable in many respects to CPM 86 having similar calling procedures to low level drivers.

The Graphical Kernal System GKS

This system appears to be the principal emerging graphics standard at programmer level, ref (12). By providing a consistent interface in high level languages such as Pascal and Fortran, GKS allows portability of graphics between different computer installations. As GKS provides a graphical interface and syntax that are common to several systems it enhances the ability of programmers to work on different systems. The programmer can access a virtual display device which may be a VDU, printer or plotter directly from the HLL. The system supports logical devices and provides access to a number of graphical primitives. It also includes features such as viewing transformations graphical input and inquiry into the system environment.

GKS supports the following functions:

A comprehensive set of drawing primitives with variable attributes.

Raster graphics area fill and pixel array primitives.

A generalised drawing primitive allowing access to the unique capability of a particular device.

Multiple workstation/display surfaces are possible through the use of logical devices. The workstation display surface may be a CRT, a plotter or a graphics printer.

Graphics primitives:

polyline - draws a straight line
polymark - creates a vector to identify points on a curve
text - displays text strings at any position with any orientation

223

fill - supports raster devices
pixel array - creates a 2D array of pixels of different colours
attributes - line type (dotted/dashed, etc) and colour
viewing and transformation - GKS provides a set of viewing transform·
ations

Graphics input functions:

request locator - returns image position in world coordinates
request valuator - returns data from device such as an A-D converter
request choice - returns an integer for a software switch
pick - returns graphic segment number that corresponds to object
being selected
request string - function reads character input from keyboard device
inquiries - the inquire capability provides information for the
applications programmer relating to system status

To add the GKS features to HHL's it will be necessary to either
respecify the language to include new reserved words associated with
graphics functions or to create ad hoc extensions to the standard
syntax. Critics of GKS argue that its lack of built-in 3D graphics
facilities is a significant omission from the standard.

4.5 THE VIRTUAL DEVICE INTERFACE VDI

The virtual device interface is standard interface between software
and graphics devices. The purpose of the VDI is to make all devices
appear as identical virtual graphics devices by defining a standard
input/output protocol. All unique characteristics of a particular
device are taken into account within the device driver software
module. The underlying strategy behind the VDI concept is to create
compatibility between different hardware configurations.

4.6 THE GRAPHICS SYSTEM EXTENSION GSX

GSX is an operating system program overlay which extends the
facilities of CPM by providing extensive graphics facilities that
satisfy the GKS standard interface. The GSX program consists of
three major components: the graphics device operating system GDOS,
the graphics input/output system GIOS, and the Gengraf utility
routine.

GDOS - this program is similar to the standard CPM BDOS, ref (4,5,6),
and contains the device independent section of GSX

GIOS - this program contains the device dependent drivers and is
analogous to the BIOS in standard CPM

Gengraf - this utility configures a graphics application program to run in the GSX environment.

The Graphics Software System Kernel GSS
GSS is a linkable runtime library module which provides an interface between the graphics application program and GSX.

Programmers Minimal Interface to Graphics PMIG
PMIG has been recently proposed as a module that will provide a subset of GKS functions for users whose applications are not graphics intensive. This interface software is primarily intended for use in small inexpensive microcomputer systems.

4.7 COST OF GRAPHICS FACILITIES

High resolution graphics facilities for microcomputers are expensive because such systems are memory intensive. A RAM area requirement of 1Mb is not untypical. As RAM memory costs fall, so will the cost of high-resolution graphics. It is important to note that the addition of a colour high resolution graphics terminal, with appropriate software, to a mono-chrome third generation microcomputer will usually be more costly than the microcomputer - up to twice the price.

4.8 NORTH AMERICAN PRESENTATION LEVEL PROTOCOL SYNTAX NAPLPS

This is a communications protocol for the transmission of two dimensional information, ref (13). By using code extension techniques six character sets of 96 characters are available to the programmer. The code extension structure is similar to the ASCII escape code sequence found in many VDU's and has been designed to support future growth in an organised manner. The character sets are outlined below. Primary character set, PCS - this includes the ASCII character set and some other well known characters.

Supplementary character set, SCS - new characters such as alpha and beta
Picture description instruction, PDI - commands to draw lines, arcs and polygons
Mosaics - these characters are similar to the British teletext characters
Macros - reduce the amount of transmitted data to and from the host terminal
Dynamically redefinable character set, DRCS - are user definable character

Two other related standards of interest are the Initial Graphics Exchange System, IGES and the Virtual Device Metafile, VDM. IGES

225

effects transfer of CAD/CAM objects, the image and its attributes, between application specific databases. VDM specifies a data format for describing a picture in a device- independent form for translation and display on a variety of graphics devices.

5.1 CONCLUSIONS

Now that small semi-portable micro-computers with graphical potential are generally available at the price of a small family car we may expect to see the increasing involvement of these machines in the day to day life of electrical engineers. As they become more powerful and more portable the usual requirement of design and development engineers working in one place becomes less and less important and data preparation will be able to be carried out at home, on site, in the works or at a customer's premises. These machines will have a common operating system and interlink with the in house machines. Care will have to be taken so that the employer will not loose the cross fertilisation of ideas occuring with centralised facilities in a design or development office but the computational paper chase we are all involved with should become a thing of the past. At the moment it is possible for a VDU/micro-computer to replace working drawings with the advantage that these could be continuously updated.

The potential of the new microcomputer will force the development of company wide scientific data bases and internationally utilised software standards. There will be a steady increase in the use of interactive analysis packages with concentration on interfacing standard units rather than complete programming exercises. The steady increase in package utilization for modelling in two or three dimensions will result in a steady conversion of these packages to the solution of design problems rather than system modelling. Finally we may even see the introduction of field based transient studies in machines to replace or complement the circuit based models in use at present.

References

1. Forghani B, Freeman E A, Lowther D A , Silvester P P -
    Interactive computer aided design of electric machines and
    electromagnetic apparatus. J. App. Phys. 53(11), Nov 82
2. Silvester P P - Interactive computer aided design in magnetics.
    IEEE Trans. on Mag. Vol Mag-17 No 6, Nov 81
3. Forghani B - Interactive modelling of magnetization curves.
    IEEE Trans. on Mag. Vol Mag-18, No 6, Nov 82
4. CPM 80 Interface Guide, Digital Research 77
5. CPM 80 Alteration Guide, Digital Research 78
6. CPM 86 Operating System Programmer's Guide, Digital Research 81

7. MSDOS Operating System Programmer's Guide, Microsoft 82
8. Tindle J - Hardware 16 bit Microprocessors, invited paper at Sirius 1 Hardware Software Symposium, 24 Feb 83
9. Ibbitson I R - Graphics, invited paper at Sirius 1 Hardware Software Symposium, 24 Feb 83
10. Ibbitson I R - Hardware/Software Interfacing, invited paper at Sirius 1 Hardware Software Symposium, 24 Feb 83
11. Tindle J - The Pascal Programming Language, invited paper at Sirius 1 Hardware Software Symposium, 24 Feb 83
12. GKS Functional Description, Draft International Standard, ISO/DIS7942, version 7.02, 9 August 82
13. NAPLPS, American Standards Institute ANSI, Doc No BSRX3 110-198X
14. Lowther D A - A Microprocessor Based Electromagnetic Field Analysis System, IEEE Trans on Mag. Vol Mag-18 No. 2 Mar 82

CPM is a Trademark of Digital Research

A Novel Approach To Computer Architecture

C Dennison
Perkin-Elmer Data Systems, Slough

Perkin-Elmer have recently introduced a new computer with unique features geared for the simulation industry. Its name is 3200 Multi-Processing System (3200 MPS). This computer was actually specified by a large simulation customer and then implemented, built and marketed by Perkin-Elmer. Before talking about the 3200 MPS let me however, give you a bit of background about Perkin-Elmer's computer history.

Perkin-Elmer produced the first 32-bit mini in August 1974. Perkin-Elmer were making 32-bit minis when other manufacturers were still producing 16-bit machines. The advantage of a 32-bit machine is of course the much higher performance capability available and the ability to run large programs, both key requirements for simulation. Perkin-Elmer has been making 32-bit machines for 10 years and we are now into our 3rd generation. From our history it is clear that Perkin-Elmer has more experience in 32-bit minicomputers than almost anybody else. The software is also fully matured with well over 900 man years of effort having been spent on it. This maturity has great benefits for the end user in terms of reliability.

Let's now turn to the 3200 MPS and see how its particular design is uniquely suited for number crunching simulation applications. The Model 3200 MPS is the top of the line in Perkin-Elmer's successful Series 3200.

It provides the user with a CPU and Auxiliary Processing Units all connected to an integrated memory and I/O system. The 3200 MPS is a multi-processing system designed for the user whose needs surpass that of one single processor and whose computation requirements are likely to grow larger.

The 3200 MPS addresses the problem that users have often faced, that is, more processing power is required than can be obtained from a single processing unit. Users have therefore been forced into multiprocessor solutions. Typically, solutions have been composed of multiple single processors, with single processor software systems. The drawback to this approach has been that there is no multi-processor awareness in the hardware or software. The user is faced with integrating the software and thus faced with developing complex coordination software. Because multi-processors are really the joining of single uni-processors, there is a high system cost associated with

this approach. With these solutions come large floor space requirements, large system costs, and high life cycle costs.

The 3200 MPS CPU is a controller of system resources. Under the
CPU's control is the I/O system, the Auxiliary Processing Units, and
of course the memory itself. It is through the CPU that the user in-
teracts with the Model 3200 MPS. All I/O intensive tasks are directe
to the CPU to realize the full benefits of the Model 3200 MPS architecture.

Providing full recognition and action to faults is also a function of
the CPU and software.

One of the key benefits of the Model 3200 MPS is PLUG-IN performance.
3200 MPS PLUG-IN performance means field expansible processing power
The system is composed of a CPU and Auxiliery Processing Units. The
CPU and auxiliary processing system is a two-cabinet system which
provides about 5 million single precision Whetstone instructions per
second. Additional Auxiliary Processing Units can be ordered, eithe
initially or field installed, and will yield an incremental processor
performance of 2 million Whetstones. Finally, a maximum of 9 Auxiliary Processing Units can be configured in the system for a total
power of 21 million Whetstones.

Just as the CPU has been optimized for system control, the Auxiliary
Processing Units have been optimized for compute-intensive applications. This optimization includes a microcode scheduler for a fast
context switch and low system overhead. Operations of the APU are
managed by the CPU, yet application tasks have communication, coordination, and synchronization capabilities with other tasks. Each
processor comes equipped with hardware floating point and writeable
control store for high performance processing.

Just as processing power can be added, so can I/O. Field-expansible
I/O means that you can grow the basic 8 port 10 megabytes-per-second
Direct Memory Access capability to a full 32 port 40 megabytes per
second. Sixty-four devices can be connected to each port. I/O
capability can be expanded to 1,023 devices.

3200 MPS also comes in compact cabinetry. The Auxiliary Processing
Units reside in their own cabinet and can be installed up to 3 per
cabinet. The full complement of memory, up to 16 megabytes, is
housed in the CPU cabinet. Power for the CPU, memory, battery back-
up and communications to the Auxiliary Processing Units is also
housed in this cabinet. Finally, APU's and I/O can reside in the
same cabinet yielding maximum utilization of available cabinet space.

The full range of the single operating system's services is available
to all tasks operating on all processors. It is the CPU that

has been optimized for control of I/O and it is therefore most logical that the CPU supports the interactive environments. These interactive environments are MTM, which is the timesharing system provided by Perkin-Elmer, and Reliance which is Perkin-Elmer's high-performance transaction processing relational database software. Since all environments are controlled by a single OS, all interactive environments can coexist comfortably with multi-processor real-time number-crunching applications.

The user has a wide choice of programming languages by which he can bring the full performance of 3200 MPS to bear. FORTRAN VII has been extended through new instructions to exploit the inter-processor facilities available on the 3200 MPS. Other scientific languages such as PASCAL and CORAL are also available.

In conclusion, it can be seen that the benefit provided by the 3200 MPS line is modularity in both hardware and software. Because of this modularity, further growth is assured. A user can also reap the benefits of low life cycle costs since he can start from a small beginning and grow his power availability. Extraneous hardware is reduced, modular programming is encouraged, and in addition, no complex processor coordination tasks are required. Additional features are added merely by adding another application task and assigning that task to an Auxiliary Processing Unit. Of course the low acquisition cost makes 3200 MPS attractive today, and the traditionally high availability and reliability associated with Perkin-Elmer systems cannot be overlooked. On all fronts, 3200 MPS is designed to provide the user with the most complete set of hardware and software tools for applications where parallel processing is beneficial so that systems will be on time, on budget, and still have room to grow easily.

231

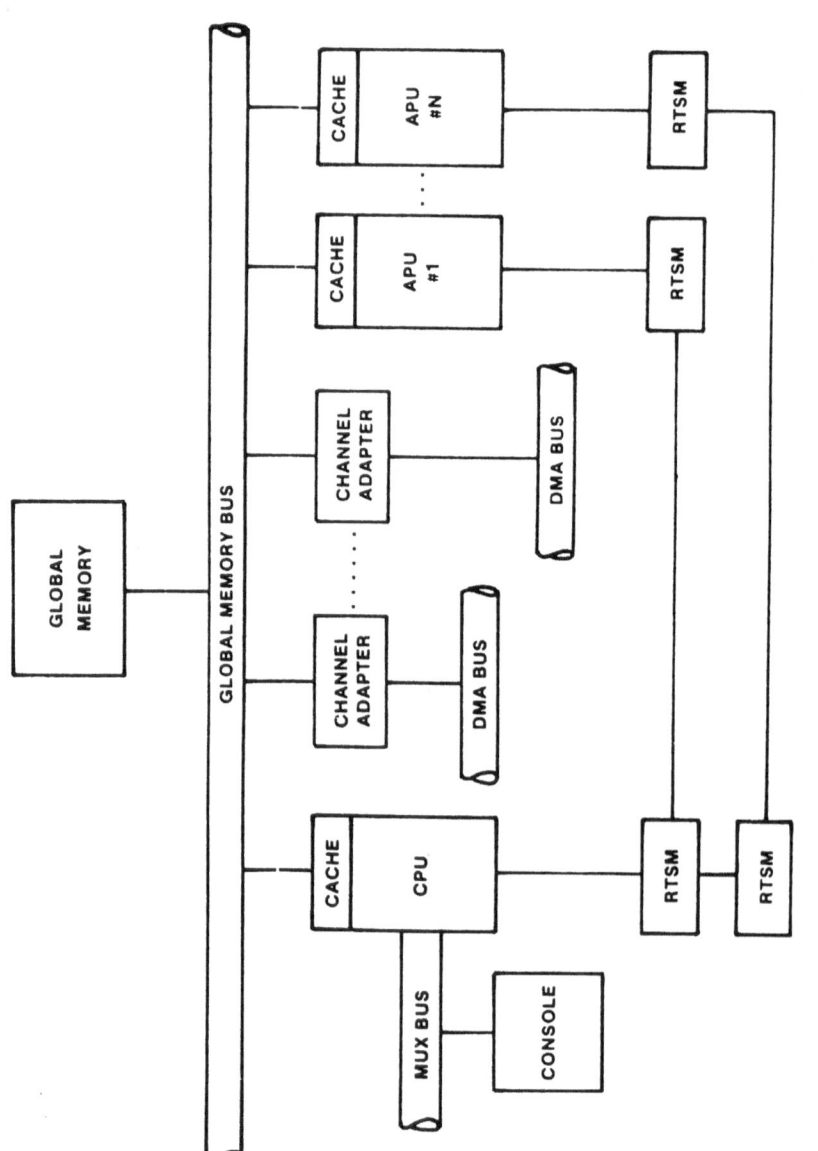

FIGURE 1. BLOCK DIAGRAM OF MODEL 3200MPS SYSTEM

LIST OF DELEGATES

Mr M H Aliabadi	Teesside Polytechnic
Dr A D Appleton	International Research & Development Co. Ltd.
Mr M L Barton	Westinghouse Electric Co, Pittsburgh
Mr C Birch	Sunderland Polytechnic
Mr R Bradley	Newcastle upon Tyne Polytechnic
Dr A Bush	Teesside Polytechnic
Dr J Caldwell	Newcastle upon Tyne Polytechnic
Professor Z J Cendes	Carnegie-Mellon University, Pittsburgh
Dr M V K Chari	General Electric Co., New York
Dr E Crompton	Sunderland Polytechnic
Dr C Dennison	Perkin-Elmer Data Systems, Slough
Dr P Dyke	Sunderland Polytechnic
Professor E M Freeman	Imperial College, London
Miss G Gregory	NEI Parsons, Newcastle upon Tyne
Dr W Hall	Teesside Polytechnic
Mr J A Hampton	Napier College, Edinburgh
Mr L Haydock	Trent Polytechnic
Dr D Howe	University of Sheffield
Mrs Hu Qing Lan	University of Cambridge
Dr A G Jack	University of Newcastle upon Tyne
Mr W F Low	University of Sheffield
Professor D A Lowther	McGill University, Montreal
Mr D Mecrow	NEI Parsons, Newcastle upon Tyne
Mr A O Moscardini	Sunderland Polytechnic
Dr S G Mudge	Wolverhampton Polytechnic
Dr M J O'Carroll	Teesside Polytechnic
Dr D O'Kelly	University of Bradford
Dr J Penman	University of Aberdeen
Dr D H Prothero	International Research & Development Co. Ltd.
Mr C W Richards	Thames Polytechnic
Dr E H Robson	Sunderland Polytechnic
Dr S Rudzinski	Central Electricity Research Lab., Leatherhead
Dr T Saleh	British Gas Corporation, Cramlington
Dr R Saunders	Sunderland Polytechnic
Mr S Scott	Newcastle upon Tyne Polytechnic
Mr G J Sears	University of Salford
Dr K Sharples	The City University
Professor P P Silvester	McGill University, Montreal
Mr A J Tindle	Sunderland Polytechnic
Mr D R Treece	NEI Parsons, Newcastle upon Tyne
Mr J F Waddington	Electricity Council Research Centre, Capenhurst
Dr A Walton	NEI Parsons, Newcastle upon Tyne

Mr M C Warnes	British Gas Corporation, Cramlington
Professor A Wexler	University of Manitoba
Dr A H Whitfield	Loughborough University
Mr T S Wilkinson	NEI Parsons, Newcastle upon Tyne
Mr S Willcock	University of Durham
Mr A Zisserman	Sunderland Polytechnic